INTERNATIONAL CENTRE FOR MECHANICAL SCIENCES

COURSES AND LECTURES No. 146

CARLO FERRARI
TECHNICAL UNIVERSITY, TURIN

LECTURES ON RADIATING GASDYNAMICS

GENERAL EQUATIONS AND BOUNDARY CONDITIONS

COURSE HELD AT THE DEPARTMENT
OF MECHANICS OF FLUIDS
OCTOBER 1972

UDINE 1974

SPRINGER-VERLAG WIEN GMBH

Originally published by Springer-Verlag Wien New York in 1972

ISBN 978-3-211-81204-4 ISBN 978-3-7091-2922-7 (eBook)

DOI 10.1007/978-3-7091-2922-7

PREFACE

This textbook is based partly on the author lectures on "Radiating Gasdynamics" delivered at the "New York University" (N.Y.) in 1966 and partly on more recent researches carried out by the same author and his group at the "politecnico" of Torino.

The author found a very exciting atmosphere in Udine at the CISM: the helpful discussions with the other Speakers and the observations or remarks of the participants to the Course have been very useful to the author, who is deeply greatful to all of them.

The author is very much indebted to Prof. Sobrero, CISM, for having given to him the opportunity to have so nice experience and for his cooperation.

Torino, 15 February, 1974 *Carlo Ferrari*

1. Introduction: Object of the lectures; essential hypothesis

The object of these lectures, is that of giving:

a) general equations of a gas flow, in which the interaction between the flow and radiation fields are taken into account:

b) the appropriate boundary conditions.

I shall also make mention rapidly of some simple applications of the theory here developed, but the applications, as well as other aspects here not considered of the radiating gasdynamics will be given by the other Speakers.

As far as the interest of these researches is concerned, I have only to mention how much they are related to them of the Astrophysics, for instance to study the phenomena that occur in the interstellar atmosphere, such that of shock wave and ionization front propagation; and the Astronautics, particularly in the reentry problem: as for this problem, it is meaningful the fig. 1. (taken by the NASA TR R-159 by Howe and Viegas [1]).

In this figure are represented by thin solid lines some possible trajectories of spacecraft with different missions; by the thick solid line the boundary of the region in which ionization occurs. Are also drawn the lines corresponding to given values of the pressure behind the shock, and the lines giving the ratio between the heat transfer rates by radiation and by convection, in according to the evaluation made by Yoshikawa and Which [2] for a nose of a 5 feet radius: this evaluation has to be considered roughly approximated, but the conclusions that can be drawn by it are probably correct. It appears clearly that, while the radiation effect is negligible for the re-entry of a satellite, the opposite occurs for space-craft re-entering from missions on the Moon, or Mars or other heavenly bodies.

The general equations will be deduced on the basis of the following fundamental hypothesis: as for the fluid, it will be considered only a gas consisting in a mixture of neutral atoms, electrons and simply ionized ions. Besides, the hypothesis of the "quasi-neutrality" for the fluid will be made: this means that it will be assumed that a separation of the electric charges will not be considered, because the Debye length, that is a measure of the distance at which the separation can occur, is much smaller than any other characteristic length of the phenomena under consideration.

This implies obviously the admission that the induced electric field is

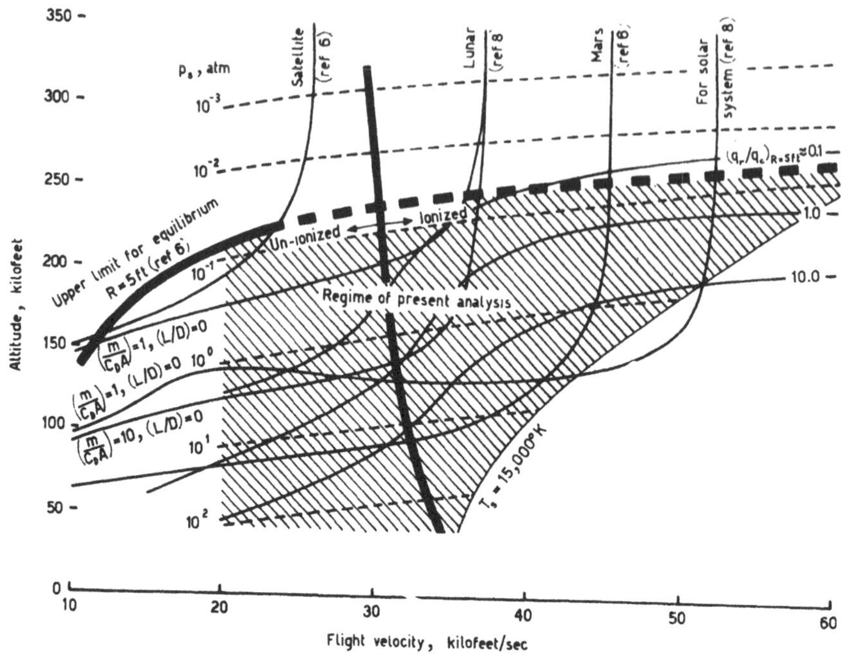

Fig.1 Flight regime

negligible, and all the electric charged particles have the same macroscopic mass averaged velocity: in addition to this admission, the stronger hypothesis will be made that all the charged and uncharged particles have the same macroscopic mass velocity. This assumption is justified by the fact that the ion-atom cross-section is much bigger than the atom-atom cross-section, and therefore the ion-atom mean free path is small compared to the atom-atom mean free path. On the contrary, it will be assumed that the electrons temperature T_e can be different from that of the other heavier particles, because their mass m_e is much smaller than that of the atoms and ion (for H, $m_e / m_a = 9.108 \times 10^{-28} / 1.673 \times 10^{-24} = 5 \times 10\text{-}4$) while it is well known that in this condition the collisions are much less efficient in the kinetic energy transfer from one of the particle to the other than in the case in which both the particles have the same, or almost the same mass. In effect, the energy transferred in a collision between one electron and one ion is, f.i., only a fraction of the order m_e / m_i i of their relative kinetic energy; consequently, the average kinetic energies of the electrons and ions can be different even for a large time. On the contrary, just because of the reason now indicated, the atoms and ions temperatures can be taken identical: this is the assumption that will be made now, as well as it will be assumed that the particles of any type are in thermodynamic equilibrium as for the translational degree of freedom, and consequently the velocity distribution is, for any species, Maxwellian. Finally it will be assumed that the atoms, and eventually also the ions, can be in any excited state.

2. Basic concepts – Atomic structure.

It is well known that any atom consists of a positively charged core, or nucleus, which contains nearly all the mass, surrounded by negatively charged, constantly moving electrons. The nucleus contains both positively charged particles, called protons, and particles of about the same mass but no charge, called neutrons. The number of the electrons is equal to the number of the prothons, it is called "atomic number", and it gives the position of each element in the periodic system of the elements.

According to the Bohr-Sommerfeld model of the atom, that is accepted at present for simplicity, the electrons travel in orbits that can be considered approximately as ellipses, having one of the focuses at the center of gravity of the nucleus; however, according to the quantum mechanics, not all the ellipses, which satisfy such a condition, are permitted but only those corresponding to certain rules, and precisely: the major and minor axis of the "permitted" ellipses are determined respectively by the quantum numbers n and ℓ, while the orientation of the plane of the orbit is determined by a third quantum number m_ℓ ; n is the "principal" quantum number, and the contribution to the energy of the atom corresponding to the electron under consideration is essentially depending on the value of n, and precisely the greater is the energy, the larger is n; ℓ is the "azimutal" quantum number and on its value is depending the moment of momentum of the electron with respect to the center of gravity of the nucleus; finally m_ℓ is the "magnetic" quantum number and on its value it is depending the component of the anular momentum of the electron in the direction of a magnetic field eventually applied to the atoms.

Finally, to complete the description of the possible configurations of an atom, it must still be observed that any electron not only moves around the nucleus, but also rotates around its axis. The angular velocity of this rotation is such that the corresponding angular momentum has an absolute value which is equal to

$|s| = 1/2\ h/2\pi$ (h is the Planck constant, $h = 6.6234 \times 10^{-27}$ erg.sec.). This rotation is called "spin", and the number $m_s = \pm 1/2$ is called "quantum number of the spin" (the double sign corresponds to the two possible senses of the spin).

The "state" of an atom depends on its electronic configuration, and this is determined giving the orbits on which the electrons move, and therefore the four quantum numbers of each electron; however, not all the possible electronic configurations are permitted, but only those that satisfy the Pauli exclusion principle: according to this principle the maximum number of the electrons that can exist in one same given orbit corresponding to given values of n, l, m_ℓ, is two, the one with positive spin ($m_s = 1/2$), the second with negative spin ($m_s = -1/2$).

Among all the "permitted" configurations there exists one corresponding to the minimum value of the energy of the atom, and the corresponding atom state is called "ground state".

In this "ground state", if the number of the electrons is greater than two, not all the electrons are equivalent, but they may be divided in inner and outer electrons. The excited states, corresponding to the greater of the atom energy, are produced, generally, when the outermost electron, which is called "optical electron", is in whatever of the highest orbits.

One may represent the various possible excited states of any atom by means of the energy-level diagram, in which each energy-level is represented by an horizontal straight line, being the vertical distance between the two parallel straight lines corresponding to two successive levels proportional to the energy difference between these levels: thus the fig. 2 gives the energy-level diagram for the H atom. In this diagram, which is only schematic, the lines corresponding to the values of n = 1; 2; 3; etc.. are signed; to $n \to \infty$ it is corresponding the orbit at infinite distance from the nucleus, and therefore the state in which the electron is completely detached.

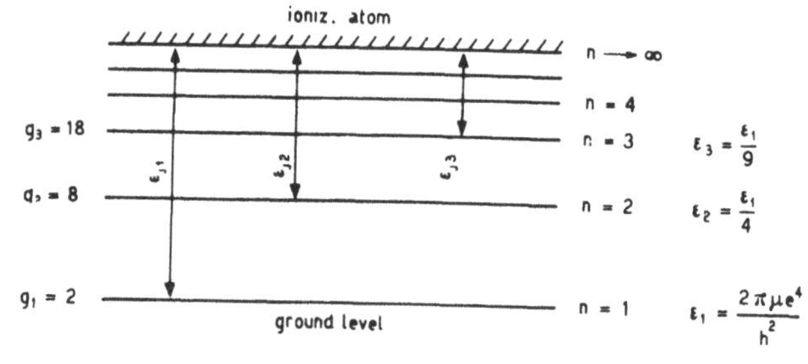

Fig.2

In this condition the atom is ionized, and the ionization energy (from the ground level), namely the energy that must be supplied to the atom in order to produce the detachment of the electron from the atom, is measured by the distance between the two lines n = 1; n = ∞. Very often the ionization energy (from any n level), and the excitation energy (from the level $n_1 < n$ to the n level) are given in electron-volt (denoted by the symbol e v): 1e v is the energy that assumes one electron, starting at rest, when it is accelerated by a potential difference equal to one volt. Since 1 volt = 1/299.729 u.e.s. and the electronic charge of one electron is $e = 4.8 \times 10^{-10}$ u.e.s., it turns out $1\ e\ v = 4.8 \times 10^{-10}/299.792 = 1.606 \times 10^{-12}$ erg.

It must be said that, since the value of the energy depends only or essentially on the value of the quantum number n, while the possible electronic configurations are depending also on the values of l, m_ℓ, and m_s, for each electron, it turns out that to a same level of energy can correspond several states: the number of the states at the same energy level is called "statistical weight" of the level itself. To complete the diagrammatic representation of the set of the energy-levels it is therefore still necessary to give for each level the corresponding statistical weight g_n (see fig. 2).

It must be observed right now that the modern quantum theory does not assign electrons to planetary orbits, but, according to the concepts of the wave mechanics, it is concerned with wave amplitude, or wave function Ψ, whose square expresses the probability of finding an electron at a given point. Although the planetary orbit must be discarded, the concept of the energy-level diagram remains intact, as well as the physical meaning of each of the four quantum number n, l, m_ℓ, m_s.

3. Atoms spectra.

Consider an atom in an excited state: it is possible to calculate the probability of the transition of this atom to another state of lower energy-level, and a basic postulate of the quantum theory is that, when an atom jumps from one state of energy ϵ_n to another state of less energy ϵ_m, spontaneously, it emits radiation in accordance with the law

$$\epsilon_{nm} = \epsilon_n - \epsilon_m = h\nu_{nm} \qquad (n > m) \qquad (3.1)$$

In this relation ϵ_{nm} is the quantum of energy (or photon) emitted by the atom; while ν_{nm} is the frequency of the radiation: thus, for the Bohr-Sommerfeld model of the H atom, the possible frequencies of the emitted radiation are

(3.2) $$\nu_{nm} = \frac{1}{h}\frac{2\pi^2\mu\ e^4}{h^2}\left(\frac{1}{m^2} - \frac{1}{n^2}\right) = \mathcal{R}_y\left(\frac{1}{m^2} - \frac{1}{n^2}\right)$$

where μ is the reduced mass of the electron ($\mu = m_a m_e/m_a + m_e \cong m_e$) if m_a and m_e are respectively the masses of the atom and of the electron); $\mathcal{R}y = 2\pi^2\ \mu e^4/h^3$ (and therefore $h\mathcal{R}y = 13.957$ e v $= 21.8368 \times 10^{-12}$ erg.) is the so called Rydberg constant.

It appears therefore that an atom is able to emit radiation in a discrete spectrum of frequencies, that are called "emission spectral lines", because of their appearance in the experimental apparatus.

It is also possible the inverse process: if the atoms are in a radiative field, there exists a certain probability that one atom, in a certain energy level m absorbs one photon $h\nu_{nm}$ and jumps from the state of energy ϵ_m to the state of energy ϵ_n, being always satisfied the rel (3.1): thus, besides the emission spectral lines, there exists the "absorption spectral lines", which can be obtained if radiations pass through a sheet of an absorbing medium, and, at the exit, are analyzed with a spectrograph.

On a photographic plate it appear light lines (the absorption spectral lines). Whatever be the atom, whatever be its complexity, the corresponding emission and absorption spectra may be more or less complex, but they are always of the same nature: all the spectral lines represent transition between two different levels; however, it must be observed right now, that not all the possible combinations of the levels are corresponding to possible transitions: certain selection rules have to be satisfied.

It has been considered so far only transitions between two discrete levels, and in this case the atom absorbs or emits precisely the correct amount of energy. But, in order to eject an electron, the atom may absorb any amount of energy equal to, or greater than that required to go from its initial state to the one represented by $n = \infty$. The excess of energy (with respect to that of ionization) imparts a velocity v to the free electron, and the equation which is satisfied, instead of the eq. (3.1) is

(3.3)

where v is the velocity of the free electron, and ϵ_{jm} stands for the ionization energy (from the level m).

This process corresponds to the photoionization of the atom. In the invers process (neutralization) one electron collides with one ion and forms a neutral atom in the state of m energy level: the excess of energy of the colliding electron with respect to the energy ϵ_m of the atom is radiated, and the frequency of the radiation is given by

$$\nu = \frac{\frac{1}{2} m_e v^2 + \epsilon_{jm}}{h} \qquad (3.4)$$

and therefore, since the kinetic energy of the free electron is not quantized, ν can take any value equal to, or greater than ϵ_{jm}/h.

Thus the total absorption spectrum, f.i. of any monoatomic gas consists of one part, called "discrete spectrum", in which there are the spectral lines corresponding to the transitions between distinct levels of energy, and a second part, the "continuous spectrum", corresponding to the photoionization, in which the radiation has any frequency greater than a given value (fig. 3). This limit frequency ν_j can be obtained by the rel. (3.4) writing in this rel. v = 0 ; it turns out that

$$\nu_j = \frac{\epsilon_{jm}}{h} . \qquad (3.5)$$

It must be observed right now, that the kind of spectrum above indicated corresponds to a schematic representation of the real phenomenon: in effect, one can never obtain a sharp line of a given frequency ν_{nm} (in other words the radiation is never monocromatic), since there are several causes that produce a broadening of the spectral lines:

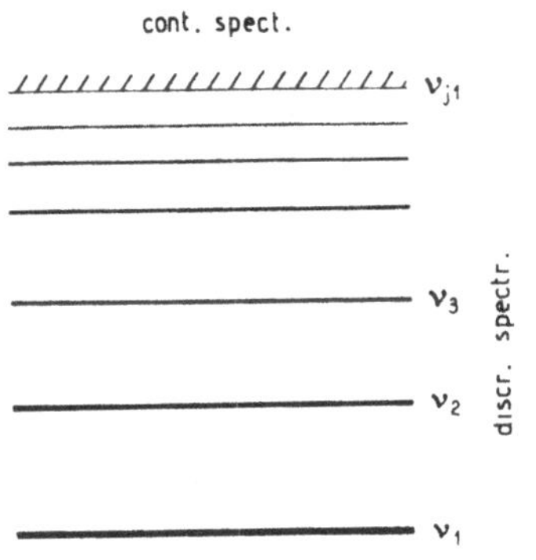

Fig.3

a) Doppler effect arising from the thermal random motions of atoms;
b) radiation damping which is a consequence of the finite lifetimes of excited levels;
c) collision damping. A radiating atom is perturbed by its neighbours (atoms, electrons, ions);
d) perturbation effect produced by electric fields (Stark effect) or by magnetic fields (Zeeman effect).

As a consequence of these perturbations, the intensity of the emitted radiation is not given by a Dirac function $\delta(\nu-\nu_{nm})$, but it has finite

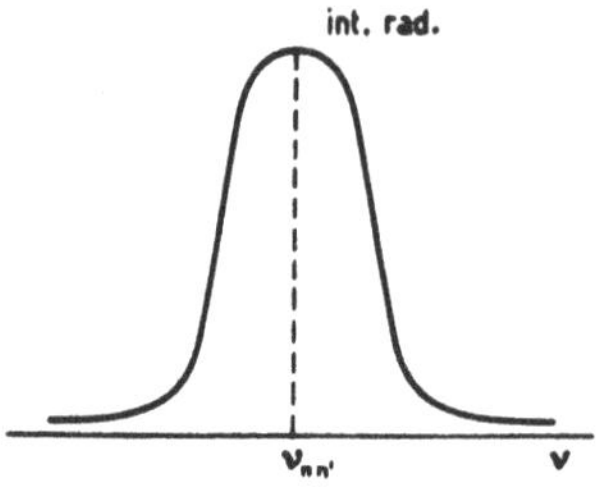

Fig.4

not null values at any value of ν : as it will be shown later on, the intensity has a very sharp maximum value at $\nu=\nu_{nm}$, and then falls very quickly to very small values when $\nu\neq\nu_{nm}$, so that the diagram, which gives the dependence of the intensity as function of the frequency has a roughly bell-shaped profile, as indicated in the fig. 4.

4. General equations: transport equations for the gas.

From what it has been indicated above it appears why interaction between flow and radiation fields occurs: the emission and absorption of radiation can produce neutralization or ionization, or, in any case, a variation of the population in the various levels of energy, and thus contribute to the transport of mass momentum and energy of each species of particles. On the other hand, as a consequence of the flow, changes of pressure, density and temperature of each species, and therefore changes of physical quantities on which the radiation field depends, occur.

The general equations of the dynamics of a radiating gas express just the balance of all contributions, included those of radiation, to the mass, momentum and energy transport for the mixture of atoms (in the various energy levels), ions, electrons, and photons.

To deduce the equations which define the rate of change in density of the material particles of any type (α) in a given internal state (i), one has to apply the Boltzmann equation to the number of density $N_{\alpha i}$ of these particles: the left-side member of the Boltzmann equation, that expresses the rate of change of the number of particles of type α in the internal state i in a given volume element, plus the variation in unit time of the same quantity due to the convection across the surface enclosing the same volume element, can be written as $\partial/\partial t\ N_{\alpha i}+\mathrm{div}(\vec{V}_{\alpha i}\ N_{\alpha i})$ being $\vec{V}_{\alpha i}$ the average mass velocity.

On the other hand, the right-side member of the Boltzmann equation, which gives the rate of change of $N_{\alpha i}$ due to the interactions of all particles of all types, can be written as sum of the contributions of the collisions between the mass particles (collisional source function) and of the interactions between these mass particles and the photons (radiative source function). Thus it may be written

$$\frac{\partial}{\partial t} N_{\alpha i} + \text{div}\,(\vec{V}_{\alpha i} N_{\alpha i}) = (M_c)_{\alpha i} + (M_r)_{\alpha i} \tag{4.1}$$

denoting with the subscript c the collisional contribution, and with the subscript r the radiative contribution.

Let be $\vec{U}$ the mass average velocity of the mixture, which is defined by

$$\rho\vec{U} = \Sigma_{\alpha i}\, m_{\alpha i}\, N_{\alpha i}\, \vec{V}_{\alpha i} \tag{4.2}$$

being the density $\rho = \Sigma_{\alpha i}\, m_{\alpha i}\, N_{\alpha i}$,and $m_{\alpha i}$ the mass of the particle α_i ; let be $\vec{u}_{\alpha i}$ the diffusion velocity of the component α_i given by

$$\vec{u}_{\alpha i} = \vec{V}_{\alpha i} - \vec{U} \tag{4.3}$$

Under the hypothesis justified in the Introduction that all the particles of any type have the same macroscopic average mass velocity it is

$$\vec{u}_{\alpha i} = 0 \tag{4.4}$$

Thus the eq. (4.1) can be written under the form

$$\frac{\partial}{\partial t} N_{\alpha i} + \text{div}(N_{\alpha i}\vec{U}) = (M_c)_{\alpha i} + (M_r)_{\alpha i} \tag{4.5}$$

Multiplying by $m_{\alpha i}$ and writing $\rho_{\alpha i} = m_{\alpha i} N_{\alpha i}$, so that $\rho_{\alpha i}$ is the density of the component α_i , one obtains

$$\frac{\partial}{\partial t} \rho_{\alpha i} + \text{div}(\rho_{\alpha i}\vec{U}) = m_{\alpha i}(M_c)_{\alpha i} + m_{\alpha i}(M_r)_{\alpha i} \tag{4.6}$$

which is the continuity equation (or equation of the conservation of mass) for the component α_i of the mixture.

Summing the eqs. (4.6) over α_i , being

$$\Sigma_{\alpha i}\,[m_{\alpha i}(M_c)_{\alpha i} + m_{\alpha i}(M_r)_{\alpha i}] = 0$$

from the requirement of mass conservation, and taking into account the rel. (4.4), one obtains

$$\frac{\partial \rho}{\partial t} + \text{div}(\rho\vec{U}) = 0 \tag{4.7}$$

which is the continuity equation of the mixture.

Following an analogous procedure it is possible to deduce the momentum equation of the mixture, which takes the form

$$\rho \frac{\partial \vec{U}_j}{\partial t} + \rho \sum_k U_k \frac{\partial U_j}{\partial x_k} = -\sum_k \frac{\partial P_{jk}}{\partial x_k} = -\sum_k \frac{\partial P_{jk}^{(t)}}{\partial x_k} - \sum_k \frac{\partial P_{jk}^{(r)}}{\partial x_k} \qquad (j,k = 1,2,3)$$

(4.8)

where the stress tensor P_{jk} has been expressed as sum of the thermal $P_{jk}^{(t)}$ and radiative $P_{jk}^{(r)}$ stress tensor. As it is well known, the components of the thermal stress tensor can be expressed by means of the rel.

$$P_{jk}^{(t)} = p^{(t)} \delta_{jk} - \tau_{jk}^{(t)}$$

being δ_{ik} the unit tensor, $p^{(t)}$ the thermodynamic pressure given by $p^{(t)} = \kappa \Sigma N_{\alpha i} T_{\alpha i}$ ($T_{\alpha i}$ = temperature of the species α_i; κ the Boltzman constant), while $t_{jk}^{(t)}$ are the components of the viscous stress tensor. Analogously, the components $P_{jk}^{(r)}$ can be written under the form

(4.9)
$$P_{jk}^{(r)} = p^{(r)} \delta_{jk} - \tau_{jk}^{(r)}$$

where the meaning and the expressions of the quantities $p^{(r)}$ and $t_{jk}^{(r)}$ will be given later on.

As for the energy equation of the mixture, one obtains

(4.10)
$$\rho \left(\frac{dH}{dt} - \frac{1}{\rho} \frac{dp}{dt} \right) = - \operatorname{div}(\vec{q} + \vec{F}) + \Phi$$

where: H is defined by the rel.

(4.11)
$$H = E + \frac{p}{\rho} = E^{(t)} + \frac{p^{(t)}}{\rho} + E^{(r)} + \frac{p^{(r)}}{\rho}$$

being E the internal energy per unit mass and p the pressure: in the rel. (4.11) both these quantities have been expressed as sum of the contributions to the same quantities corresponding to the thermal motion and to the radiation. The thermal internal energy $E^{(t)}$ can be obtained by the rel.

(4.12)
$$E^{(t)} = \frac{1}{\rho} \Sigma_{\alpha i} N_{\alpha i} \left(\frac{3}{2} \kappa T_{\alpha i} + \epsilon_{\alpha i} \right)$$

whereas the internal energy $\epsilon_{\alpha i}$ corresponding to the electronic excitation of the atoms and ions has to be measured from the same energy level: thus, for instance, taking equal to zero the internal energy of the atoms at the ground state, the internal energy of ions, singly ionized, (corresponding to the internal degrees of freedom), at the ground level, is equal to the ionization energy ϵ_j from the ground level; the Boltzmann constant κ has the value $K = 1.38 \times 10^{-16}$ erg / °K

The vector $\vec{q}$ is the vector energy flux due to the thermal motion of the mass particles, while $\vec{F}$ is the radiative energy flux; Φ is the energy dissipation function, that, when it is written in the usual form $\Phi = \Sigma_{j\,k} \tau_{jk}(\partial U_j / \partial x_k)$ can also be represented as sum of the part due to the thermal motion, and of the part corresponding to the radiation.

Besides the energy equation of the mixture one has still to consider the energy equation for the electrons alone, since their temperature T_e has been assumed different from that T of the heavier particles. This equation can be written under the form

$$\frac{\partial}{\partial t}\left[\rho_e\left(E_e + \frac{1}{2}U^2\right)\right] + \operatorname{div}\left\{\left[\rho_e\left(E_e + \frac{1}{2}U^2\right) + p_e\right]\vec{U} + \vec{q}_e\right\} = Q_e^{(c)} + Q_e^{(r)} \qquad (4.13)$$

if one neglects the energy dissipation due to the viscous stresses, since the corresponding viscosity coefficients are very small (of the order of magnitude of those of the neutral gas reduced in the ratio m_e / m_a); E_e is the thermal internal energy of the electrons and therefore $E_e = (1/\rho_e)[N_e (3/2)\kappa T_e]$ is the thermal energy flux due to the thermal conductivity of the electrons.

The right-side member of the eq. (4.13) has the menaing of "energy source" for the electrons; this source appears to be separated in the two parts, the one corresponding to the collisions (elastic and anelastic), the other corresponding to the radiation.

5. Radiative energy transfer equation.

The equation of radiative energy transfer can be obtained emphasizing the particle aspect of radiation and considering therefore the radiation field composed by $n^{(r)}$ photons per unit volume.

Now, the velocity and the momentum of a photon with frequency ν, travelling in the direction of the unit vector $\vec{s}$, are

$$\vec{v} = \vec{s}c \; ; \; \vec{p}_\nu = \vec{s}\,\frac{h\nu}{c} \qquad (5.1)$$

if c is the light speed. Let be $f_\nu(\vec{s};\vec{x})$ the number of photons in the momentum state $\vec{p}_\nu$ (or, in other words, travelling in direction $\vec{s}$). The forces acting on the photons, arising from the external region with respect to the region in which radiation is passing, are null or completely negligible, and therefore the Boltzmann equation for photons has the form

$$\frac{\partial}{\partial t} f_\nu(\vec{s};\vec{x}) + c\,\vec{s}\,\mathrm{grad}\, f_\nu(\vec{s};\vec{x}) = W_\nu(\vec{s};\vec{x}) \tag{5.2}$$

In this equation the first term at the left side gives the rate of increase of the number of photons in momentum state $\vec{p}_\nu(\vec{s})$ in a fixed volume element, while the second term represents the corresponding variation in unit time due to the "convection" of photons across the surfaces limiting the volume element above indicated. The member at the right side represents the net rate of increase of the photons under consideration as a result of the interaction between photons and particles in the same volume element.

Now, there are two kinds of such interactions:

a) interactions that do not produce any change of the energy of the photons; these are roughly analogous to the elastic collisions between molecules of one chemical species with molecules of other species, and they are called "scattering" In other words, photons are "scattered" by the atoms of a gas when only the direction of propagation of the photons is changed. Now, in most situations of interest in gasdynamics scattering is unimportant and therefore it will be not considered here.

b) interactions in which the energy of the photons is changed: they are those corresponding to the phenomena of absorption and emission, about which mention has been already made, and that will be considered with much more details later on. It is, however, necessary to observe right now, that, as for the emission, another phenomenon occurs in addition to that already considered.

Suppose an atom is in the excited level n, higher than the level n′: it has been already said that there exists a not null, finite probability that in the unit time the atom will spontaneously (i.e. without any external agency) jump from level n to level n′ with the emission of energy $h\nu_{nn'} = \epsilon_n - \epsilon_{n'}$. If however radiation of frequency $\nu_{nn'}$, corresponding to the permitted transition falls upon the atom, the likelihood of its cascading to level n′ with the emission of $\nu_{nn'}$ radiation is increased by an amount dependent upon the intensity of the incident light, or, in other words, upon the number of the photons interacting with the atom in the unit time.

This phenomenon, which corresponds to a negative absorption, is called "induced emission", and it must still be observed that the induced radiation is put back in the beam in the same direction as the incident radiation, and not emitted in a random direction as are the spontaneously emitted quanta: it is this property, which is utilized in the gasdynamic "laser".

It turns out from what above has been said, that one can formally write

$$W_\nu(\vec{s};\vec{x}) = \left[\frac{\partial}{\partial t} f_\nu(\vec{s};\vec{x})\right]_{e,a,s} \tag{5.3}$$

where the subscripts e, a, s stand for the emission, absorption, scattering contributions. When scattering is neglected, one can write

$$W_\nu(\vec{s};\vec{x}) = c\Sigma_a\left\{-N_a\sigma_{ab}(\nu)f_\nu(\vec{s}) + N_b\sigma_{ba}(\nu)[1+f_\nu(\vec{s})]\right\} \tag{5.4}$$

The term proportional to N_a represents absorption, and the term proportional to N_b represents total emission, i.e. spontaneous emission plus induced emission; N_a is the number of density of absorbing particles in energy state a; summation over a means summation over all states that can have transitions to another energy level by absorption of photons of frequency ν. Analogously, N_b is the number of systems in the upper energy state b, defined in similar way: a particle goes from state "a" to state "b" when a photon of frequency ν is absorbed, and, conversely, a particle goes from state "b" to state "a" when a photon of frequency ν is emitted. The quantities σ_{ab}, σ_{ba} have the physical dimensions of an area, and in effect, as it is apparent from the expression (5.4) they are the cross-sections for the corresponding processes of absorption and emission.

Now, in quantum theory it is found that the number of momentum states per unit volume between $\vec{p}_\nu$ and $\vec{p}_\nu + d\vec{p}_\nu$ is given by $h^{-3}p_\nu^2\,dp_\nu d\Omega$ whereas $d\Omega$ is the elemental solid angle centered about the direction of $\vec{p}_\nu$ (i.e. $\vec{s}$). Thus the number of photons per unit volume with frequencies between ν and $\nu + d\nu$ and travelling within the elemental solid angle $d\Omega$ about $\vec{s}$ is $(2/c^3)f_\nu(\vec{s})\nu^2 d\nu d\Omega$ if one takes in account the relations (5.1), while the factor 2 arises from the summation over photon polarization.

Multiplying such a number by $h\nu$, energy of one photon, one obtains

$$\frac{2h\nu}{c^3} f_\nu(\vec{s};\vec{x})\nu^2 d\nu\, d\Omega \tag{5.5}$$

which gives the energy per unit volume of the radiation in the direction $\vec{s}$ and in the frequency range $\nu\div\nu+d\nu$. On the other hand, let be $I_\nu(\vec{s};\vec{x})$ the specific intensity of the radiation in a given point $(\vec{x})$ and in a given direction $\vec{s}$, so defined: $I_\nu(\vec{s};\vec{x})$ is the amount of radiative energy, which flows per second, in unit solid angle, in unit frequency interval, through a unit area placed perpendicularly to the direction $\vec{s}$. It may be deduced that the same amount of energy given by the rel. (5.5) can be expressed as $(I_\nu/c)d\nu d\Omega$, and therefore it turns out that

$$I_\nu(\vec{s};\vec{x}) = \frac{2h\nu^3}{c^2} f_\nu(\vec{s};\vec{x}) \tag{5.5'}$$

If one takes $W_\nu(\vec{s})$ from the rel. (5.4), and expresses f_ν by means of I_ν using the rel. (5.5'), one deduces

$$\frac{1}{c}\frac{\partial}{\partial t} I_\nu(\vec{s}) + \vec{s}\cdot \text{grad } I_\nu(\vec{s}) = \sum_a \left\{ - N_a \sigma_{ab} I_\nu(\vec{s}) + N_b \sigma_{ba}\left[\frac{2h\nu^3}{c^2} + I_\nu(\vec{s})\right]\right\} =$$

$$= -\left\{\sum_a (N_a\sigma_{ab} - N_b\sigma_{ba})\left[I_\nu - \frac{\sum_a N_b\sigma_{ba}}{\sum_a (N_a\sigma_{ab} - N_b\sigma_{ba})}\frac{2h\nu^3}{c^2}\right]\right\} \tag{5.6}$$

Consider now a system (C_e) composed by the same particles of the one (C) under consideration, but in complete local thermodynamic equilibrium: these words mean that the distribution of atoms among the energy levels (included the continuum) corresponds to the Boltzmann value (or Saha equation) for the gas temperature assumed to be equal to the electron temperature T_e (it is quite obvious that if free electrons do not exist, and therefore all the components of the mixture are atoms in the various energy level with the same temperature T, the gas temperature has to be taken equal to T). Since C_e is in equilibrium, the rate of change of any quantity, and therefore of $I_\nu(\vec{s})$, has to be null, and this condition has the consequence that the second factor within the [ ] at the right side of the rel. (5.6) is null. On the other hand, since the radiation for the system C_e is in equilibrium with the matter, the specific intensity reduces to the Planck function $B_\nu(T_e)$ given by the rel.

$$B_\nu(T_e) = \frac{2h\nu^3}{c^2}\left[\exp(h\nu/kT_e) - 1\right]^{-1} \tag{5.7}$$

while the number of density N_a and N_b have to have the values corresponding to the condition of complete local thermodynamic equilibrium (for $T = T_e$): they will be denoted by the symbols $\bar{N}_a$ and $\bar{N}_b$.

Then, it must be observed that the cross-section σ_{ab} corresponding to the interactions between photons and atoms, or ions, is an atomic constant, and therefore its value is not depending if the system is or not in equilibrium: this property will be shown with more details later on, but it is opportune to say right now that it will be indicated later that it exists a case in which the same property is not satisfied.

On the contrary, the emission from the particle in energy state b may depend on the number of density of other particles, when this emission is a result of the interaction between these particles and particles (b): this case occurs, for instance, in the process of recapture of a free electron by an ion with spontaneous and induced emission of radiative energy.

Since the number of density depends on the state of the entire system, it appears that σ_{ba} may be depending on this state: let be $\bar{\sigma}_{ba}$ the value of σ_{ba} corresponding to the system C_e associated to C in any point P. Thus for C_e one can write

$$B_\nu = \frac{\Sigma_a \bar{N}_b \bar{\sigma}_{ba}}{\Sigma_a (\bar{N}_a \sigma_{ab} - \bar{N}_b \bar{\sigma}_{ba})} \frac{2h\nu^3}{c^2} \qquad (5.8)$$

Solving for $2h\nu^3/c^2$, and substituting in (5.6) gives

$$\frac{1}{c}\frac{\partial}{\partial t} I_\nu(\vec{s}) + \vec{s}\cdot \text{grad}\, I_\nu(\vec{s}) = -k_\nu (I_\nu - \varphi B_\nu) \qquad (5.9)$$

whereas

$$k_\nu = \Sigma_a (N_a \sigma_{ab} - N_b \sigma_{ba}) \; ; \; \varphi = \frac{\dfrac{\Sigma_a \bar{N}_a \sigma_{ab}}{\Sigma_a \bar{N}_b \bar{\sigma}_{ba}} - 1}{\dfrac{\Sigma_a N_a \sigma_{ab}}{\Sigma_a N_b \sigma_{ba}} - 1} \; . \qquad (5.10)$$

It is clear from the relation (5.9) that k_ν represents the absorption coefficient per unit volume of the system under consideration, while the function φ takes the value one when the state of complete thermodynamic equilibrium exists, and therefore φ represents an index of the disequilibrium of the system.

The eq. (5.9) is the equation of radiative energy transfer in its general form: now it must be observed that the first term at the left side of this eq. is very small with respect to the other terms, at least in a great variety of problems interesting the Gasdynamics, and therefore from now forward it will be neglected.

6. Cross-sections for the interactions between photons and material particles.

The eq. (5.9) is basic for the dynamics of the radiating gases, because it defines the variation law of the specific intensity I_ν and, on the other hand, it will be shown that all the other quantities, which give the radiative contribution to the transport of mass of the different species, momentum and energy, can be expressed by means of I_ν. It requires, however, that the expressions of the cross-section σ_{ab} and σ_{ba} for the interaction phenomena between radiation and material particles be determined: now a large number of interactions with all these particles is possible, but, fortunately, in most of the problems interesting the Gasdynamics the interactions that are "dominant" for the phenomena to study are in number rather limited. These interactions will be considered separately, and first of all there will be investigated those corresponding to the processes in which no free electrons are involved, i.e. the electrons are bound to the nucleus both before and after the interaction (bound-bound processes); there will be considered later on the processes in which free electrons exist (bound-free; free-free processes).

6.a Interaction in the bound-bound processes.

As it has been already said, there exists a not null, finite probability that an atom in an excited state n jumps spontaneously in the unit time from the level n' to a lower level n . Let be A (n,n') such a probability: the coefficient $A(n, n')$, which is usually written as $A_{nn'}$. is called the Einstein coefficient of the spontaneous emission. The total number of downward $(n \to n')$ transitions in a time dt will be therefore $N_n A_{nn'} dt$ if, as usual, N_n is the number of atoms, per unit volume, in level n. On the other hand, if the atom is exposed to radiation, it can absorb energy from the radiation field, and therefore transition from the level n to a higher level n" will take place at a rate proportional to the number of the atoms in the level n, N_n. and to the intensity $I_{\nu''}$, of the radiation of frequency $\nu(n, n'')$: the total number of transitions $(n \to n'')$ is given by $N_n B_n(n,n'') I_{\nu''}$, where the coefficient $B(n, n'')$, that usually is written as $B_{nn''}$ is called Einstein coefficient of absorption.

In addition to these phenomena there is still the phenomenon of induced emission, as it has been observed at No. 5: thus, one has to define another Einstein coefficient, the coefficient of negative absorption, or induced emission $B_{nn'}$ $(n' < n)$ such that the number of induced emissions of atoms from the level n to the level n' per unit time will be

$$N_n B_{nn'} I_{\nu'}$$

The total number of atoms leaving n for n' in a time dt will be

$$N_n (A_{nn'} + B_{nn'} I_{\nu'})dt.$$

It is to observe right now that the three coefficients $A_{nn'}$, $B_{n'n}$, $B_{nn'}$ are not independent, but they are connected by two relations that can be deduced considering the system in the thermodynamic equilibrium: under this condition the intensity I_ν is given by the Planck functions B_ν

$$I_\nu = B_\nu = \frac{2h\nu^3}{c^2} [\exp(h\nu/\kappa T) - 1]^{-1} \quad (6.1)$$

while the relative population of the two levels n and n' is given by the Boltzmann formula:

$$\frac{\bar{N}_n}{\bar{N}_{n'}} = \frac{g_n}{g_{n'}} \exp(-\epsilon_{nn'}/\kappa T) \quad (6.2)$$

where $\epsilon_{nn'} = \epsilon_n - \epsilon_{n'}$ is the energy difference between the two levels n and n'. and g_n $g_{n'}$ are the statistical weights of the two levels.

Express the condition that in thermodynamic equilibrium the number of transitions from n to n' is equal to the number of transitions from n' to n, so that

$$\bar{N}_n (A_{nn'} + B_{nn'} B_\nu) = \bar{N}_{n'} B_{n'n} B_\nu \;;$$

then consider that the Einstein coefficients are atomic quantities, that depend on the properties of the particular atom under consideration, and not on the statistical properties of the ensemble to which this particular atom belongs, so that the A and B coefficients must be independent on the temperature T. Thus one obtains the relations

$$\frac{B_{nn'}}{B_{n'n}} = \frac{g_{n'}}{g_n} \;; \quad \frac{A_{nn'}}{B_{n'n}} = \frac{2h\nu^3}{c^2} \frac{g_{n'}}{g_n} \quad (6.3)$$

from which one has

$$\frac{A_{nn'}}{B_{nn'}} = \frac{2h\nu^3}{c^2} \,. \quad (6.4)$$

Although the rels (6.3) have been obtained considering conditions of thermodynamic equilibrium, it is important to realize that the same relations will hold under all conditions, just for the reason above indicated of the independence of the values of the coefficients A and B on the properties of the whole system of atoms.

Consider now an atom in a level n. It can go spontaneously to certain lower levels n'; the rate at which the atoms in the upper level descend spontaneously to the lower states is proportional to the sum of the A values of the corresponding transitions, according to the definition of the Einstein coefficient A: thus one can write:

(6.5)
$$\frac{dN_n}{dt} = - N_n \Sigma_{n'} A_{nn'} = - N_n \Gamma_n ,$$

(6.6)
$$\Gamma_n = \Sigma_{n'} A_{nn'} .$$

By integration of (6.5) one obtains

(6.7)
$$N_n = N_n^{(o)} \exp\,(-\Gamma_n t)$$

where $N_n^{(o)}$ denotes the number of the atoms in level n at $t = o$. It appears therefore that Γ_n is simply the reciprocal of the mean lifetime T_n of an atom in level n, i.e.

(6.8)
$$T_n = \frac{1}{\Gamma_n} = \frac{1}{\Sigma_{n'} A_{nn'}}$$

from which it appears another physical meaning of the $A_{nn'}$-coefficients. The quantity Γ_n is called "quantum radiative damping coefficient" according to the rel.(6.7).

In the deduction now indicated it has been neglected the fact that the lower levels themselves may have finite lifetime.

A more exact treatment of the problems shows that while eq. (6.8) is valid for the resonance line (namely, the line corresponding to the transition from the first excited state to the ground state), in case of transitions between levels n and n' that are both excited, the damping constant $\Gamma_{nn'}$ is related to T_n by the rel.

(6.8')
$$\Gamma_{nn'} = \frac{1}{T_n} + \frac{1}{T_{n'}} = \Gamma_n + \Gamma_{n'}$$

or it is equal to the sum of the reciprocal of the lifetime of the upper and lower levels.

It must be observed, just now, that until now it has been supposed that an atom at a given level n has a well defined energy ϵ_n; but, in effect, this is not true, because it follows from the uncertainty principle of Heisenberg that, if $T_{n'}$ is the mean lifetime of an atom in a given level n' there is an uncertainty $\Delta\epsilon_{n'}$ of the energy of the atom such that

$$T_{n'} \, \Delta\epsilon_{n'} = \frac{h}{2\pi} \, . \tag{6.9}$$

Therefore any excited state has a natural breadth which varies inversely as the lifetime in that state. Now, Weisskopf and Wigner [3] have shown that the proportion of atoms in a broadened state of mean energy $\epsilon_{n'}$, having energies within ϵ to $\epsilon + d\epsilon$ is given by the probability law

$$W_n(\epsilon)\, d\epsilon = \frac{(\Gamma_{n'}/h)\, d\epsilon}{\left(\frac{2\pi}{h}\right)^2 \left(\epsilon - \epsilon_{n'}\right)^2 + \left(\frac{1}{2}\,\Gamma_{n'}\right)^2} \tag{6.10}$$

where $\Gamma_{n'}$ denotes again the reciprocal of the lifetime of the state under consideration. For an excited state, $\Gamma_{n'}$ is of the order of magnitude 10^{-8} sec. and therefore it is apparent from the rel. (6.10) that the function at the right side has a very sharp maximum when $\epsilon = \epsilon_{n'}$, so that most of the atoms have energies near the mean energy, and only a very small fraction $(dN_{n'}/N_{n'})$ on the $N_{n'}$ atoms in the state n' have energies within ϵ to $\epsilon + d\epsilon$ with $\epsilon \neq \epsilon_{n'}$. Considering a broadened energy-level n higher than n', and assuming that the absorption and emission of radiation occur in a system of atoms containing two and only two broadened energy states, between which transitions are possible, Weisskopf and Wigner in [3] deduced as expression of the absorption coefficient per unit volume, corresponding to the transition $(n' \to n)$ the relation

$$k'_\nu = \frac{1}{4\pi} B_{n'n} \frac{h\nu_{nn'}}{\pi} \frac{\delta}{(\nu - \nu_{nn'})^2 + \delta^2} N_{n'} \tag{6.11}$$

where

$$\nu_{nn'} = \frac{\epsilon_n - \epsilon_{n'}}{h}$$
$$\delta = \frac{\Gamma_n}{4\pi} + \frac{\Gamma_{n'}}{4\pi} = \frac{\Gamma_{nn'}}{4\pi} \, . \tag{6.12}$$

and therefore δ is proportional (according to the constant of proportionality $(4\pi)^{-1}$) to the sum of the reciprocals of the lifetime of an atom in the levels n and n'. Thus one obtains

$$(6.13) \qquad \sigma_{n'n} = \frac{\kappa'_\nu}{N_{n'}} = \frac{1}{4\pi} B_{n'n} \frac{h\nu_{nn'}}{\pi} \frac{\delta}{(\nu - \nu_{nn'})^2 + \delta^2}$$

On the other hand it is deduced by the quantum theory that the Einstein coefficient $B_{n'n}$ can be written under the form

$$(6.14) \qquad B_{n'n} = \frac{4\pi^2 e^2}{m_e c} f_{n'n} \frac{1}{h\nu_{nn'}}$$

where the coefficient $f_{n'n}$ is called "oscillator strength": the physical meaning of such coefficient, which explains the denomination given to it, is that the absorption due to one quantum atom is equivalent to that produced by f classical (or Herz) oscillators. The $f_{n'n}$ coefficients are typical for any atom: thus the f values for H may be computed by the formula (see [4])

$$(6.15) \qquad f_{n'n} = \frac{2^6}{3\sqrt{3}\,\pi} \frac{1}{2n'^2} \frac{1}{\left(\frac{1}{n'^2} - \frac{1}{n^2}\right)^3} \frac{1}{n^3 n'^3} .$$

The problem of the determination of the f values for He cannot be solved esplicitly, and resort must be had to approximation techniques such as the variation method, which Goldberg has employed for the most important transitions in this atom [5]; in many of complex atoms accurate f values can not be calculated by theoretical means and recourse must be had to experimental determinations. As function of $f_{n'n}$ it is also possible to express the quantum damping constant Γ_n, using the relation between $B_{n'n}$ and $A_{nn'}$; thus one obtains

$$(6.16) \qquad A_{nn'} = \frac{g_{n'}}{g_n} \frac{8\pi^2 e^2 \nu_{nn'} f_{n'n}}{m_e c^3} .$$

It turns out that

$$(6.17) \qquad \Gamma_n = \Sigma_{n'} A_{nn'} = \frac{8\pi^2 e^2}{m_e c^3} \Sigma_{n'} \nu^2_{nn'} \frac{g_{n'}}{g_n} f_{n'n} .$$

If one substitutes the rel. (6.14) into the rel (6.13) one obtains

$$\sigma_{n'n} = \frac{e^2}{m_e c} f_{n'n} \frac{\delta}{(\nu - \nu_{nn'})^2 + \delta^2} \tag{6.18}$$

where δ has to be evaluated by means of the rels. (6.12) (6.17) and the one (analogous to (6.17), corresponding to $\Gamma_{n'}$.

The expression of $\sigma_{n'n}$ now deduced takes into account the natural broadening of the spectral lines, which is due only to the radiative damping; however it has been already observed that there are many other interactions which influence such a broadening. If one considers, beside the radiative damping, the effect of the collisions, one obtains a formula for $\sigma_{n'n}$ which is identical to that given by (6.18): δ , however, is no more equal to $\Gamma_{m'}/4\pi$ solely, but it is defined at least approximately, by

$$\delta = \frac{\Gamma_{n'n}}{4\pi} + \frac{2}{T_o}\frac{1}{4\pi} \tag{6.19}$$

where To is the mean value of the time interval between two collisions capable of producing certain critical effect (precisely a given phase shift of the waves emitted by the atom represented as a classical oscillator). A theory has been developed by Weisskopf and Lorentz [6]; a more detailed discussion of the problem has been given by Lindholm [7].

The problem becomes more complicated if one takes into account also the Doppler effect: roughly speaking, one may say that the Doppler effect modifies the absorption cross-section $\sigma_{n'n}$ to give an error curve in the core of the line (i.e. for $\nu \cong \nu_{nn'}$),without change in the wings (i.e. for $\nu \gtrless \nu_{nn'}$)[8]. By means of the expression of $\sigma_{n'n}$ one can deduce the corresponding expression of $\sigma_{nn'}$. It must be observed, first of all, that in the bound-bound process under consideration, not only $\sigma_{n'n}$ but also $\sigma_{nn'}$ is an atomic constant, because it is concerned with an interaction process between one atom and photon alone, so that its value does not depend on the statistical properties of the system, and therefore it is

$$\sigma_{nn'} = \bar{\sigma}_{nn'} \; ; \quad \bar{\sigma}_{n'n} = \sigma_{n'n} \tag{6.20}$$

On the other hand, considering a system in thermodynamic equilibrium at the temperature Te, in which the only transition that takes place is the one from n' to n and viceversa, one finds that

$$(6.21)\qquad \frac{2h\nu^3_{nn'}}{c^2}\,\frac{\bar{N}_n\sigma_{nn'}}{\bar{N}_{n'}\sigma_{n'n}-\bar{N}_n\sigma_{nn'}} = \frac{2h\nu^3_{nn'}}{c^2}\left[\exp(h\nu_{nn'}/\kappa T_e)-1\right]^{-1}$$

where it has been assumed that in the very small interval of the values of ν in which $\sigma_{n'n}$ is sensibly different from zero, one can take the value of any function of ν equal to that corresponding to $\nu=\nu_{nn'}$; therefore it turns out that

$$(6.22)\qquad \frac{\bar{N}_n}{\bar{N}_{n'}}\,\frac{\sigma_{nn'}}{\sigma_{n'n}} = \exp(-h\nu_{nn'}/\kappa T_e)$$

Now the ratio $\bar{N}_n/\bar{N}_{n'}$ is given by the Boltzmann distribution law

$$(6.23)\qquad \frac{\bar{N}_n}{\bar{N}_{n'}} = \frac{g_n}{g_{n'}}\exp(-h\nu_{nn'}/\kappa T_e)$$

so that one deduces

$$(6.24)\qquad \frac{\sigma_{nn'}}{\sigma_{n'n}} = \frac{g_{n'}}{g_n}$$

6.b. Interaction in the bound-free processes.

The bound-free processes are those in which either the initial or the final state, but not both, includes a free electron, In the case of absorption, since the atom loses one electron and becomes ionized, these processes are also called "photoionization processes"; they are represented by the notation

$$(6.25)\qquad A_n + h\nu \rightleftarrows A^+ + e + \left(\frac{1}{2}m_e v^2\right)$$

where the last term$(1/2\, m_e v^2)$indicates that the excess of the energy of the photon with respect to the energy necessary to ionize the atom A in the n energy level appears as kinetic energy of the liberated electron.

Since this kinetic energy is not quantized, the frequency ν of the photon, which is absorbed, can have any value greater than a certain limit value, or threshold: this limit value can be obtained easily by observing that the energy eq. is

$$(6.26)\qquad h\nu = \epsilon_{jn} + \frac{1}{2}m_e v^2$$

where ϵ_{jn} is the ionization energy of the atom at energy level n; thus setting v = 0 one obtains that the lowest value (ν_{jn}) of ν is given by

$$\nu_{jn} = \frac{\epsilon_{jn}}{h} . \tag{6.27}$$

Let be considered at first the H atom; for this atom it has been deduced

$$\epsilon_{jn} = \frac{h\mathcal{R}_y}{n^2} \tag{6.28}$$

being $\mathcal{R}_y$ the Rydberg constant, while in the transition($n \to n''$; $n'' > n$) it is

$$\nu_{nn''} = \mathcal{R}_y\left(\frac{1}{n^2} - \frac{1}{n''^2}\right) \tag{6.29}$$

Now Menzel and Pekeris [9] suggested that the rel. (6.29) be applied also to the transitions involving the continuum (namely, transitions involving frequencies of any value ν greater than ν_{jn}) by means of the substitution $n'' = i\chi$, where χ is any real number, not necessarily an integral number. Thus for these transitions it results

$$\nu = \mathcal{R}_y\left(\frac{1}{n^2} + \frac{1}{\chi^2}\right) \tag{6.30}$$

from which, by comparison with (6.26) one finds

$$\frac{1}{2} m_e v^2 = \frac{h\mathcal{R}_y}{\chi^2} \tag{6.31}$$

Consider the continuum portion of the spectrum adjacent to a given frequency ν^* as the limit configuration of the one consisting of lines corresponding to the values of χ given by $\chi = \chi^* + \eta x$, whereas χ^* is the value of χ for $\nu = \nu^*$ and η is a very small quantity, while x is any integral number, when η is tending to zero; use for the transition ($n \to \chi$) the expression of the cross-section given by the rel. (6.18) and deduce the cross-section of the photoionization process as the limit expression of $\sigma_{n\chi}$ for $\eta \to 0$; one finds

$$\sigma_{nj} = \frac{32}{3\sqrt{3}} \frac{\pi^2 e^6}{ch^3} \frac{\mathcal{R}_y}{n^5 \nu^3} \tag{6.32}$$

The (6.32) is known as Kramer's formula.

A more accurate deduction of the cross section σ_{aj}, by the quantum theory leads to multiply the right side member of the rel. (6.32) by a correction factor, which is called Gaunt factor and is given by the formula

(6.33) $$g' = 1 - 0.173\left(\frac{h\nu}{\epsilon_{jH}}\right)^{1/3}\left[\frac{2}{n^2}\left(\frac{\epsilon_{jH}}{h\nu}\right) - 1\right]$$

being ϵ_{jH} the ionization energy for the H atom at ground state. In most of the cases of practical interest this factor is very close to unity so that it can be neglected.

The Kramer's formula is valid only for H atoms: however, for one-electron transitions in the field of a nucleus with electric charge corresponding to the atomic number Z, plus a core of Z_b tightly bound electrons, the same relation can be applied with a sufficient accuracy for other atoms than the H atoms, if the quantum numbers n and l of the bound state are large, and if the right side of (6.32) is multiplied by Z_e^4, being Z_e, the effective atomic number, given by $Z_e = Z - Z_b$. These atoms are called "hydrogen-like atoms", and these words mean that the outermost electrons move in a Coulomb field, in which the portion $(Z_b e)$ of the electric charge of the nucleus is fully schielded by the Z_b innermost electrons. For small n and l numbers the bound electron core does not fully shield the nucleus and the relation above given falls down; if the actual energy ϵ_{jn} is known, one can take

(6.34) $$Z_e^2 = \frac{\epsilon_{jn}\, n^2}{h\mathcal{R}_y} .$$

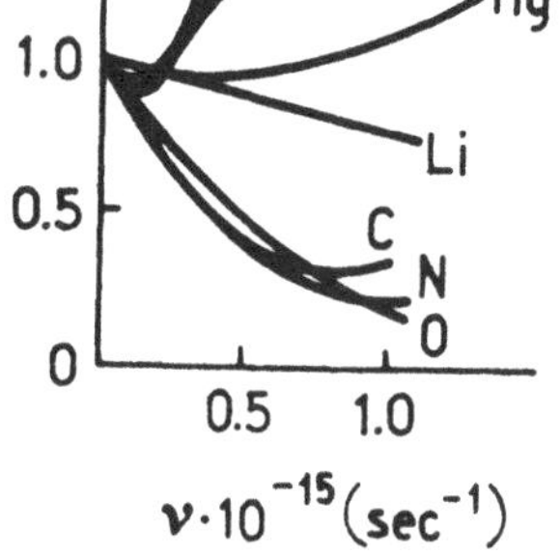

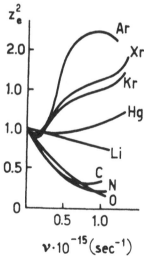

Fig.5

Besides, the right-side member of (6.32) has to be multiplied by γ/P_0, being γ equal to the ratio between the number of the sublevels of the atom under consideration corresponding to given vlaues of n and l, and the analogous quantity of the H atom; P_0 is the partition function of the atom $(P_0 = \Sigma_i g_i \exp(-\epsilon_j/\kappa T))$.

Bates and Damgaard [10] have obtained more refined approximations for non hydrogen wave functions. Biberman and Norman [11] have deduced a relationship for the σ_{nj} cross-section, which is formally identical to that above indicated, in which, however, Z_e^2 is a complex function of the frequency ν and temperature T, and varies irregularly from atom to atom. This function is shown by the diagrams of the fig. 5, which corresponds to the several atoms indicated in the same fig..

By the expression of the cross-section of the photoionization process, can be deduced the expression of the cross-section of the invers process, or radiative recombination, in analogous way as the one followed in the case of bound-bound processes.

One can observe, first of all, that in the recombination process represented by the notation

$$A^{+} + e + \left(\frac{1}{2} m_e v^2\right) = A_n + h\nu \tag{6.35}$$

the number of recombinations, in the frequency range from ν to $\nu + d\nu$ for the radiation emitted, is proportional to the product of the numbers of density of electrons and ions, and therefore the cross-section σ_{jn}, corresponding to the emission, is proportional to the number of density of the electrons, so that one can write

$$\frac{\bar{\sigma}_{jn}}{\sigma_{jn}} = \frac{\bar{N}_j}{N_j} \tag{6.36}$$

On the other hand for the system in thermodynamic equilibrium at the temperature Te one has the relation, analogous to the rel. (6.21):

$$\frac{2h\nu^3}{c^2}\left[\exp(h\nu/\kappa T_e) - 1\right]^{-1} = \frac{\bar{N}_j \bar{\sigma}_{jn}}{\bar{N}_n \sigma_{nj} - \bar{N}_j \bar{\sigma}_{jn}} \frac{2h\nu^3}{c^2} \tag{6.37}$$

from which

$$\frac{\bar{N}_j \bar{\sigma}_{jn}}{\bar{N}_n \sigma_{nj}} = \exp(-h\nu/\kappa T_e) \tag{6.38}$$

and therefore

$$\frac{\bar{\sigma}_{jn}}{\sigma_{nj}} = \frac{\bar{N}_n}{\bar{N}_j} \exp(-h\nu/\kappa T_e) \tag{6.39}$$

Thus it turns out that

$$\frac{\sigma_{jn}}{\sigma_{nj}} = \frac{\bar{N}_n}{\bar{N}_j^2} N_j \exp(-h\nu/\kappa T_e) \tag{6.40}$$

where $\bar{N}_j^2/\bar{N}_n$ is given by the Saha equation

$$\frac{\bar{N}_j^2}{\bar{N}_n} = \frac{2g_j}{g_n} \frac{(2\pi m_e \kappa T_e)^{3/2}}{h^3} \exp\left(-\frac{T_{jn}}{T_e}\right)$$

being $\epsilon_{jn} = \kappa T_{jn}$; g_j and g_n are respectively the statistical weight of the ion and the atom (at the n level).

6.c. The interaction in the free-free processes.

The free-free processes are those in which the electron is free both before and after the interaction with an atom or an ion: using again the concept of orbit, along which the electron either free or bound moves, it is well known that the geometric parameters which define the orbit of the electron around f.i. the ion are depending on the energy E of the system (electron + ion) and on the angular momentum of the relative motion.

If the energy E decreases in an anelastic collision, while the angular momentum remains constant, as it must be because the only forces acting on the colliding particles are internal, the free electron initially moving in a hyperbolic orbit, (to which it corresponds a value $E > o$), goes with continuity form this orbit to a parabolic orbit ($E = o$), then to an elliptic orbit ($E < o$): in this last case the electron remains bound to the nucleus and the radiative recombination occurs. In the intermediate cases between $E > o$ and $E = o$, the electron remains free, but it will move, after the collision, along a less energetic orbit, and the excess of the initial energy with respect to the final energy is emitted by the electron as radiative energy: the phenomenon now indicated is called "Bremsstrahlung", and it may occur either with emission of photons as in the case above described, or with absorption of photons, in which case the electron goes, obviously, to a more energetic orbit. In any case the equation of the energy, that substitutes eq. (6.26), is in the process now considered

$$h\nu = \frac{1}{2} m_e \left| v_2^2 - v_1^2 \right| \tag{6.41}$$

if v_2 and v_1 , are respectively the asymptotics velocities of the electron after and before the collision.

Following an approximate procedure Kramer deduced for the cross-section in the process of absorption by Bremsstrahlung the expression

$$\sigma_{ab} = \frac{4}{3}\left(\frac{2\pi}{3m_e \kappa T_e}\right)^{1/2} \frac{e^6}{hcm_e \nu^3} N_j \tag{6.42}$$

and from this relation it is apparent that in this type of interaction not only the σ_{ba} but also the σ_{ab} is depending on the statistical properties of the system, represented by the temperature of the electrons T_e and by the number of density of the ions N_j . On the other hand that this fact occurs it is quite obvious, since the transitions corresponding to the Bremsstrahlung are represented by the notation

$$A^+ + e + \frac{1}{2} m_e v_1^2 \rightleftarrows A^+ + e + \frac{1}{2} m_e v_2^2$$

One can immediately see that the number of the transitions (either direct or inverse) occuring in the unit volume, in the unit time, is proportional to the product of the numbers of density of ions and electrons.

The rel. (6.42), considering the procedure by means of which it has been deduced, can be considered valid, strictly speaking, only for the H atom; however, since the electric field, in which the free electron moves, is that induced by the charge of the nucleus and those of the electrons bound to the nucleus, the same formula can be applied also to more complex atoms in that one can admit that the orbit of the free electron be such that the distances of any of its points from the nucleus and from the bound electrons are equal.

The determination of the cross-section σ_{ba} for the inverse process of emission by Bremsstrahlung can be obtained by the same method above applied. It is apparent from what above has been said, that

$$\frac{\sigma_{ab}}{\bar{\sigma}_{ab}} = \frac{N_j}{\bar{N}_j} \quad ; \quad \frac{\sigma_{ba}}{\bar{\sigma}_{ba}} = \frac{N_j}{\bar{N}_j} \qquad (6.43)$$

while in conditions of thermodynamic equilibrium it must be

$$\frac{2h\nu^3}{c^2}\left[\exp(h\nu/\kappa T_e) - 1\right]^{-1} = \frac{\bar{N}_j\,\bar{\sigma}_{ba}}{\bar{N}_j\,\bar{\sigma}_{ab} - \bar{N}_j\,\bar{\sigma}_{ba}}\;\frac{2h\nu^3}{c^2} \qquad (6.44)$$

and therefore

$$\frac{\bar{\sigma}_{ba}}{\bar{\sigma}_{ab}} = \exp(-h\nu/\kappa T_e) = \frac{\sigma_{ba}}{\sigma_{ab}} \quad ; \qquad (6.45)$$

thus

$$\sigma_{ba} = \sigma_{ab}\exp(-h\nu/\kappa T_e)\,. \qquad (6.46)$$

7. The formal solution of the equation of radiative energy transfer.

When the expressions of the absorption and emission cross-sections have been specified, a formal solution of the eq. (5.8) can be easily obtained if the term $(1/c)(\partial/\partial t) I_\nu$ is assumed to be negligible: observe, however, that this is not essential assumption, and it is made only for the sake of greater simplicity of the treatment.

Thus the eq. (5.9) can be written under the form

$$\frac{\partial I_\nu}{\partial s} = \kappa_\nu \, (I_\nu - \varphi B_\nu) \tag{7.1}$$

being ds an elementary length in the direction of the unit vector $\vec{s}$.

Let be

$$d\tau_\nu = \kappa_\nu ds, \quad (7.2) \qquad \text{or} \qquad \tau_\nu = \int_{s_0}^{s} \kappa_\nu ds \quad (7.2')$$

where s_0 is the value of s corresponding to the point P_0 intersect on a surface, limiting the radiation field, by the straight-line in the direction of $\vec{s}$ through the "influenced" point P. It is apparent from these relations that τ is a dimensionless coordinate; it is called "optical thickness" and its physical meaning is the following: if one writes $d\tau_\nu = ds/l_\nu$, where $l_\nu = 1/\kappa_\nu$, it is easy to see that l_ν is the "penetration length" of the radiation, namely it is a measure of the length through which the photons can penetrate in the absorbing medium. The greater is l_ν, the greater is this length, and therefore l_ν can be interpreted as the mean free path of the photons of frequency ν; thus τ_ν is the s coordinate along the $(P_{oi}\ \vec{s})$ ray measured in photons mean free path unity.

Now, the solution of the eq. (7.1), which in the point P_0 takes a given value $I_\nu(P_{oi}\ \vec{s})$, while for $s \to \infty$; $\varphi = 1$, satisfies the condition $I_\nu = B_\nu$, is

$$I_\nu = \int_{(\tau_\nu)_0}^{\tau_\nu} \varphi B_\nu \exp[-(\tau_\nu - \tau'_\nu)] \, d\tau'_\nu + I_\nu[(\tau_\nu)_0] \exp[-(\tau_\nu - (\tau_\nu)_0)] \tag{7.3}$$

8. Radiative contribution to the transport of momentum, energy and mass.

As it has been said, one can express in terms of $I_\nu(\vec{s};\vec{x})$ the quantities which give the radiative contributions to the transport of mass of each species of the mixtures, momentum and energy.

Let us consider at first the transport of momentum: to any photon it is associated a momentum $(h\nu/c)$ in the direction of the propagation of the photon, as it has been already said and applied at n. 5.

Now, the radiative stress tensor can be obtained by the net rate of transfer of momentum through a unit area of arbitrary orientation at any general point following a method quite similar to that for the gas molecular stress-tensor. In effect, the amount of radiative energy passing per second through the unit area perpendicular to the x_i axis, within the solid angle $d\Omega$ centered about the direction of the unit vectors $\vec{s}$, whose components are l_j, is $l_j \int_0^\infty d\Omega\, I_\nu d\nu = l_j\, d\Omega\, I$, denoting with I the specific intensity integrated over all frequencies. The associated rate of transfer of momentum is $(I/c) l_i d\Omega\, \vec{s}$ and the component of this vector along the x_j axis will be therefore $(I/c) l_i l_j d\Omega$. The correspondent component of the radiative stress-tensor will be given by

$$P_{ij}^{(r)} = \frac{1}{c} \int_{4\pi} \ell_i \ell_j I\, d\Omega \tag{8.1}$$

The components $P_{ij}^{(r)}$ can be written under the form

$$P_{ij}^{(r)} = p^{(r)} \delta_{ij} - \tau_{ij}^{(r)} \tag{8.2}$$

being δ_{ij} the unit tensor;

$$p^{(r)} = \frac{1}{3c} \int_{4\pi} I d\Omega = \frac{4\pi}{3c} J \tag{8.3}$$

where J represents the intensity of the radiation averaged over all solid angle ($J = (1/4\pi) \int_{4\pi} I\, d\Omega$);

$$\tau_{ij}^{(r)} = -\frac{1}{c} (1 - \delta_{ij}) \int_{4\pi} \ell_i \ell_j I d\Omega + \frac{1}{3c} \delta_{ij} \int_{4\pi} (\Sigma'_k \ell_k^2 - 2 \ell_i \ell_j) I\, d\Omega \tag{8.4}$$

where the apex in the symbol of summation means that the summation is extended over all $k \neq i (i=j)$. One can see from the rels. (8.3) (8.4) that $p^{(r)}$ one third of the trace of the tensor $P_{ij}^{(r)}$, namely it has a value equal to the mean value of the normal tensions acting on three surface elements perpendicular each other, so that it has the meaning of pressure of radiations: the tensor $\tau_{jk}^{(r)}$ has a null trace and its components for $k \neq j$ represents the radiative tangential stresses.

In a way which is so simple as that which leads to the determination of $P_{ij}^{(r)}$ it is possible to express the radiative energy per unit volume and the contribution of radiation to the transport of the energy.

Thus, defining the density of the radiative energy the amount of electromagnetic energy which is contained in a given instant in the unit volume, one deduces

(8.5) $$n = \frac{1}{c}\int_{4\pi} I d\Omega = \frac{1}{c} 4\pi J$$

while, if one considers the radiation in the range frequency from ν to $\nu+d\nu$, the correspondent energy density u_ν is

(8.5') $$u_\nu = \frac{1}{c}\int_{4\pi} I_\nu \, d\Omega = \frac{4\pi}{c} J_\nu$$

As for the flux vector of the radiative energy $\vec{F}$ one obtains

(8.6) $$\vec{F} = \int_0^\infty \vec{F}_\nu d\nu \quad ; \quad \vec{F}_\nu = \int_{4\pi} I_\nu (\vec{s}) \, \vec{s} \, d\Omega$$

More complex is the determination of the radiative contribution to the transport of mass for each species of the mixture under consideration, namely the determination of the "radiative source function", which appears in the eq. (3.5).

Let consider the radiative processes leading to a change of the number of density N_i of the atoms at the state i and represented by the notations

(8.7) $$\left| \begin{array}{l} A_i + h\nu_{ji} \rightleftarrows A_j \quad ; \quad A_i + h\nu \rightleftarrows A^+ + e + \left(\frac{1}{2} m_e v^2\right) \\ A_k + h\nu_{ik} \rightleftarrows A_i \end{array} \right.$$

being

$$\nu_{ji} = \frac{\epsilon_j - \epsilon_i}{h} \quad ; \quad \nu_{ik} = \frac{\epsilon_i - \epsilon_k}{h} \qquad (k<i<j)$$

Now, the number of transitions per unit volume and per unit time, from the level i to the level j due to the absorption of radiation whose specific intensity is I_ν, in the range of frequencies from ν to $\nu+d\nu$, within the solid angle $d\Omega$, centered about the direction $\vec{s}$ of I_ν is given by

$$C_{\nu_{ji}} N_i B_{ij} \frac{1}{\pi} \frac{\dfrac{\Gamma}{4\pi\nu_{ji}}}{\left(\dfrac{\nu}{\nu_{ji}} - 1\right)^2 + \left(\dfrac{\Gamma}{4\pi\nu_{ji}}\right)^2} I_\nu \frac{d\Omega}{4\pi} d\frac{\nu}{\nu_{ji}} = C_{\nu_{ij}} N_i B_{ij} \Phi_\nu I_\nu \frac{d\Omega}{4\pi} d\frac{\nu}{\nu_{ji}}$$

(8.8)

being $C_{\nu_{ji}}$ defined by the rel.

$$C_{\nu_{ji}} = \frac{B_{\nu_{ji}}}{\int_0^\infty \Phi_\nu B_\nu \, d\frac{\nu}{\nu_{ji}}} \quad ; \tag{8.9}$$

here Φ_ν is the "profile" of the absorption line around the frequency ν_{ji} ; the assumption is made that this profile is the Lorentzian profile as indicated by the rel. (6.11), but the formulae that will be obtained are valid for any profile. One deduces that the total number of transitions (j→i) produced by absorption of radiation for any $\vec{s}$ and for any ν is

$$N_i B_{ij} C_{\nu_{ji}} \int_0^\infty \Phi_\nu J_\nu \, d\frac{\nu}{\nu_{ji}}$$

On the other hand the number of transitions (j→i) per unit volume and per unit time is

$$N_j \left[A_{ji} + B_{ji} C_{\nu_{ji}} \int_0^\infty \Phi_\nu J_\nu \, d\frac{\nu}{\nu_{ji}} \right]$$

where all the symbols have the meaning already given. It turns out that the contribution to the rate of change of N_i, which corresponds to the process represented by the first of the rels. (8.7) is

$$(M_r)_{ij} = - N_i C_{\nu_{ji}} B_{ij} \int_0^\infty \Phi_\nu J_\nu d\nu/\nu_{ji} + N_j \left[A_{ji} + C_{\nu_{ji}} B_{ji} \times \int_0^\infty \Phi_\nu J_\nu d\nu/\nu_{ji} \right] =$$

$$= - N_i B_{ij} \left(1 - \frac{N_j}{N_i} \frac{B_{ji}}{B_{ij}} \right) \left[C_{\nu_{ji}} \int_0^\infty \Phi_\nu J_\nu d\nu/\nu_{ji} - \frac{N_j}{N_i} \frac{A_{ji}}{B_{ij}} \frac{1}{1 - \frac{N_j}{N_i} \frac{B_{ji}}{B_{ij}}} \right]$$

Using the relations already obtained

$$\frac{B_{ji}}{B_{ij}} = \frac{g_i}{g_j} \quad ; \quad \frac{A_{ji}}{B_{ij}} = \frac{2h\nu_{ji}^3}{c^2} \frac{g_i}{g_j} = B_{\nu_{ji}} \left[\exp(h\nu_{ji}/\kappa T) - 1 \right] \frac{g_i}{g_j}$$

one obtains

$$(M_r)_{ij} = - N_i B_{ij} \left(1 - \frac{N_j}{N_i} \frac{g_i}{g_j} \right) \left[C_{\nu_{ji}} \int_0^\infty \Phi_\nu J_\nu d\nu/\nu_{ji} - \right.$$

$$- \frac{\exp(h\nu_{ji}/\kappa T) - 1}{(N_i/N_j)(g_i/g_j) - 1} B_{\nu_{ji}}(T) \Big] \tag{8.10}$$

Since it is

$$\frac{\bar{N}_j}{\bar{N}_i} = \frac{g_j}{g_i} \exp(-\epsilon_{ji}/\kappa T)$$

One can write

$$(M_r)_{ij} = - N_i B_{ij}\left(1 - \frac{N_j}{N_i}\frac{g_i}{g_j}\right)\left[C_{\nu_{ji}} \int_0^\infty \Phi_\nu J_\nu \, d\nu/\nu_{ji} - \frac{1 - \exp(-h\nu_{ji}/\kappa T)}{\frac{N_i}{N_j}\frac{\bar{N}_j}{\bar{N}_i} - \exp(-h\nu_{ji}/\kappa T)} B_{\nu_{ji}}\right] \tag{8.10'}$$

In analogous way, if $(M_r)_{ki}$ is the "radiative source function" corresponding to the transition (K→i) one finds

$$(M_r)_{ki} = - N_k B_{ki}\left(1 - \frac{N_i}{N_k}\frac{g_k}{g_i}\right)\left[C_{\nu_{ik}} \int_0^\infty \Phi_\nu J_\nu \, d\nu/\nu_{ik} - \frac{1 - \exp(-h\nu_{ik}/\kappa T)}{\frac{N_k}{N_i}\frac{\bar{N}_i}{\bar{N}_k} - \exp(-h\nu_{ik}/\kappa T)} B_{\nu_{ik}}\right] \tag{8.11}$$

being now

$$C_{\nu_{ik}} = B_{\nu_{ik}} \Big/ \int_0^\infty \Phi_\nu B_\nu \, d\nu/\nu_{ik}$$

Therefore the contribution to the "radiative source function" which corresponds to all the excitation processes represented by the first and the third of the notations (8.7) can be expressed by means of the formula

$$(M_r^{(i)})_{exc} = \Sigma_j (M_r)_{ij} - \Sigma_k (M_r)_{ki} \tag{8.12}$$

where the first summation is extended over all the values j for which $\epsilon_j > \epsilon_i$, and the second summation is extended over all values of k for which $\epsilon_k < \epsilon_i$. Finally, the contribution to M_r due to ionization process can be obtained by the following method.

The net radiative energy, which is absorbed per unit volume and in the unit time, in the frequencies range from ν to $\nu + d\nu$, within the elemental solid angle $d\Omega$ is given in the process under consideration by the relation

$$I_\nu\, d\Omega\, d\nu\, (N_i \sigma_{ij} - N_j \sigma_{ji}) - N_j \sigma_{ji} \frac{2h\nu^3}{c^2}\, d\Omega\, d\nu = N_i \sigma_{ij} \left[1 - \frac{N_j^2}{\bar{N}_j^2} \frac{\bar{N}_i}{N_i} \exp(-h\nu \kappa T_e)\right]$$

$$\cdot \left[I_\nu - \frac{1 - \exp(-h\nu/\kappa T_e)}{\frac{\bar{N}_j^2}{N_j^2}\frac{N_i}{\bar{N}_i} - \exp(-h\nu/\kappa T_e)} B_\nu(T_e)\right] d\nu \qquad (8.13)$$

and therefore the net radiative energy absorbed in the interval $\nu \div \nu + d\nu$ and corresponding to all direction of propagation $\vec{s}$ is given by

$$N_i \sigma_{ij} \left[1 - \frac{N_j^2}{N_i} \frac{\bar{N}_i}{\bar{N}_j^2} \exp(-h\nu/\kappa T_e)\right]\left[J_\nu - \frac{1 - \exp(-h\nu/\kappa T_e)}{\frac{\bar{N}_j^2}{\bar{N}_i}\frac{N_i}{N_j^2} - \exp(-h\nu/\kappa T_e)} B_\nu(T_e)\right] d\nu \qquad (8.13')$$

The correspondent number of absorbed photons, and therefore the correspondent number of photoionization, is

$$4\pi N_i \sigma_{ij} (h\nu)^{-1} \left[1 - \frac{N_j^2}{N_i} \frac{\bar{N}_i}{\bar{N}_j^2} \exp(-h\nu/\kappa T_e)\right]\left[J - \frac{1 - \exp(-h\nu/\kappa T_e)}{\frac{\bar{N}_j^2}{\bar{N}_i}\frac{N_i}{N_j^2} - \exp(-h\nu/\kappa T_e)} B_\nu(T_e)\right] d\nu \qquad (8.14)$$

The "radiative source function" which corresponds to the ionization process (from the i level) is obtained integrating the expression (8.14) over ν, so that one deduces

$$(M_r^{(i)})_{ion} = -4\pi N_i \int_{\nu_{ji}}^{\infty} (h\nu)^{-1} \sigma_{ij} \left[1 - \frac{N_j^2}{N_i} \frac{\bar{N}_i}{\bar{N}_j^2} \exp(-h\nu/\kappa T_e)\right] \cdot \left[J_\nu - \frac{1 - \exp(-h\nu/\kappa T_e)}{\frac{\bar{N}_j^2}{\bar{N}_i}\frac{N_i}{N_j^2} - \exp(-h\nu/\kappa T_e)} B_\nu(T_e)\right] d\nu \qquad (8.15)$$

To complete the discussion of the argument, it is still opportune to give also the expression of the "collisional source function" for the collisional process which correspond to the radiative processes (8.7), and are represented by the notations

(8.16) $$A_i + B \rightleftarrows A_j + B \; ; \quad A_i + B \rightleftarrows A^+ + e + B \; ; \quad A_k + B \rightleftarrows A_i + B$$

where B represents the particle colliding in the two-body on three-body collisions above indicated, and whose state is assumed not changed by the collision. The expression of the source $(M^{(i)})_c$ can be written under the form

(8.17) $$(M^{(i)})_c = \Sigma_k (N_k \chi_{ki} - N_i \chi_{ik}) - \Sigma_j (N_i \chi_{ij} - N_j \chi_{ji})$$

The quantities χ_{ki} and χ_{ij} are the reaction rates of the forward processes $(k \to i), (i \to j)$, while the quantities χ_{ik} , χ_{ij} are the reaction rates for the rearward processes $(i \to k), (j \to i)$. All these quantities have the physical dimensions of a reciprocal of a time, so that one can write

(8.18) $$\chi_{ij} = \frac{1}{\tau_{ij}} \; ; \quad \chi_{ji} = \frac{1}{\tau_{ji}} \; ;$$

where τ_{ij}, τ_{ji} , ... represent the characteristic time of excitation (or disexcitation), ionization (or neutralization), by collision. Now it is

$$\Sigma_k (N_k \chi_{ki} - N_i \chi_{ik}) - \Sigma_j (N_i \chi_{ij} - N_j \chi_{ji}) = \Sigma_k \left[N_k \chi_{ki} \left(1 - \frac{N_i}{N_k} \frac{\chi_{ik}}{\chi_{ki}} \right) \right] - $$
$$- \Sigma_j \left[N_i \chi_{ij} \left(1 - \frac{N_j}{N_i} \frac{\chi_{ji}}{\chi_{ij}} \right) \right]$$

while the number of excitation $(k \to i)$ per unit time and unit volume is given by the formula

(8.19) $$N_{k \to i} = N_k \Sigma_B N_B \int_{\epsilon_{ik}}^{\infty} v \, f^*(\epsilon) (\sigma_B)_{ki} \, d\epsilon$$

if N_b is the number of density of the particles B and the summation is extended over all species that can collide with A_k $(B \doteq A_k$ included); v is the relative velocity of the colliding particles, ϵ kinetic energy of this relative motion; $f^*(\epsilon)$ is the distribution function of this kinetic energy, while $(\sigma_B)_{ki}$ is the cross-section which corresponds to the collision between A_k and B leading to the transition $k \to i$; $\epsilon_{ik} = \epsilon_i - \epsilon_k$.

It is therefore

$$\chi_{ki} = \Sigma_B N_B \int_{\epsilon_{ik}}^{\infty} v\, f^*(\epsilon)\,(\sigma_B)_{ki}\, d\epsilon \qquad (8.20)$$

Under the assumption that the traslational degrees of freedom are in equilibrium, and the distribution function for the kinetic energy are all Maxwellian, corresponding to a same value of the temperature, if B is not an electron, while, in the case B is an electron the relative kinetic energy can be taken equal to the kinetic energy of the electron, it turns out that

$$\chi_{ki} = \Sigma_B (\chi_{ki})_B = \Sigma_B N_B \left(\frac{8\kappa T_B}{\pi\mu_B}\right)^{1/2} \int_{\epsilon_{ik}/\kappa T_B}^{\infty} z\, e^{-z}\, (\sigma_B)_{ki}(\epsilon_i\, \epsilon_{ik})\, dz = \Sigma_B (\bar{\chi}_{ki})_B \frac{N_B}{\bar{N}_B}$$

(8.21)

whereas $T_B = T$, if B is not an electron; $T_B = T_e$ if B is an electron; μ_B is the reduced mass of the colliding particles A_k and B, namely

$$\mu_B = \frac{m_B m_{Ak}}{m_B + m_{Ak}}; \quad z = \frac{\epsilon}{\kappa T_B} ;$$

$$(\bar{\chi}_{ki})_B = \bar{N}_B \left(\frac{8\kappa T_B}{\pi\mu_B}\right)^{1/2} \int_{\epsilon_{ik}/\kappa T_B}^{\infty} z\, e^{-z}\, (\sigma_B)_{ki}(\epsilon_i ; \epsilon_{ik})\, dz \; . \qquad (8.21')$$

The symbols with a bar have the same meaning as the symbols without bar, but in condition of thermodynamic equilibrium.

Now, the transition (k→i) is an excitation process, and therefore the inverse process involves also a two-body collision, so that a relation analogous to (8.21) is valid for the χ_{ik} :

$$\chi_{ik} = \Sigma_B N_B \frac{(\bar{\chi}_{ik})_B}{\bar{N}_B} \qquad (8.22)$$

and consequently one can write

$$1 - \frac{N_i}{N_k}\frac{\chi_{ik}}{\chi_{ki}} = 1 - \frac{N_i}{N_k} \frac{\Sigma_B N_B \dfrac{(\bar{\chi}_{ik})_B}{\bar{N}_B}}{\Sigma_B N_B \left(\dfrac{(\bar{\chi}_{ki})_B}{\bar{N}_B}\right)} \qquad (8.23)$$

But in local thermodynamic equilibrium it must be

$$\bar{N}_k (\bar{\chi}_{ki})_B = \bar{N}_i (\bar{\chi}_{ik})_B$$

from which it tutns out that

$$(\bar{\chi}_{ik})_B = \frac{\bar{N}_k}{\bar{N}_i} (\bar{\chi}_{ki})_B$$

Substituting this relation in the eq. (8.23) gives

$$1 - \frac{N_i}{N_k}\frac{\chi_{ik}}{\chi_{ki}} = 1 - \frac{N_i}{N_k}\frac{\bar{N}_k}{\bar{N}_i} \tag{8.23'}$$

where the relative population ($\bar{N}_k \bar{N}_i$) of the atom at the states (k) and (i) is given by the Boltzmann formula.

The transition (i→j) may be either an excitation process, or an ionization process: in the first case one has again

$$1 - \frac{N_j}{N_i}\frac{\chi_{ji}}{\chi_{ij}} = 1 - \frac{N_j}{N_i}\frac{\Sigma_B \frac{N_B}{\bar{N}_B}(\bar{\chi}_{ji})_B}{\Sigma_B \frac{N_B}{\bar{N}_B}(\bar{\chi}_{ij})_B} = 1 - \frac{N_j}{N_i}\frac{\bar{N}_i}{\bar{N}_j} \tag{8.24}$$

In the second case the recombination process involves a three-body collision and therefore it is

$$(\chi_{ji})_B = N_e N_B \frac{(\bar{\chi}_{ji})_B}{\bar{N}_e \bar{N}_B} \;;\quad \chi_{ji} = \frac{N_e}{\bar{N}_e}\Sigma_B \frac{N_B}{\bar{N}_B}(\bar{\chi}_{ji})_B = \frac{N_j}{\bar{N}_j}\Sigma_B \frac{N_B}{\bar{N}_B}(\bar{\chi}_{ji})_B \,,$$

(8.25)

being now $N_e = N_j$. One can deduce

$$1 - \frac{N_j}{N_i}\frac{\chi_{ji}}{\chi_{ij}} = 1 - \frac{N_j^2}{N_i \bar{N}_j}\frac{\Sigma_B (\bar{\chi}_{ji})_B \frac{N_B}{\bar{N}_B}}{\Sigma_B (\bar{\chi}_{ij})_B \frac{N_B}{\bar{N}_B}} \tag{8.26}$$

In local thermodynamic equilibrium it is

$\bar{N}_i (\bar{\chi}_{ij})_B = \bar{N}_j (\bar{\chi}_{ji})_B$ and therefore $(\bar{\chi}_{ji})_B = \dfrac{\bar{N}_i}{\bar{N}_j} (\bar{\chi}_{ij})_B$

thus the rel. (8.26) gives

$$1- \frac{N_j}{N_i}\frac{\chi_{ji}}{\chi_{ij}} = 1- \frac{N_j^2}{N_i}\frac{\bar{N}_i}{\bar{N}_j^2} \qquad (8.26')$$

and the collisional source function can be expressed by the eq. (8.17)

$$(M^{(i)})_c = \Sigma_k\left[N_k \chi_{ki}\left(1- \frac{N_i}{N_k}\frac{\bar{N}_k}{\bar{N}_i}\right)\right] - \Sigma_j \left[N_i \chi_{ij}\left(1- \frac{N_j^2}{N_i}\frac{\bar{N}_i}{\bar{N}_j^2}\right)\right]$$

As for the cross-sections $(\sigma_B)_{ki}$ one observes that, if B is an electron, $(\sigma_B)_{ki}$ can be obtained by the Mott, Massey and Fowler formula

$$(\sigma_{ki})_{el.} (\epsilon) = 3 f_{ki} \frac{\pi e^4}{\epsilon}\left(\frac{1}{\epsilon_{ik}} - \frac{1}{\epsilon}\right) \qquad (8.27)$$

being f_{ki} the "oscillator strength" associated to the transition ($k \to i$); while $\epsilon_{ik} = \epsilon_i - \epsilon_k$.

If the process ($i \to j$) is also an excitation process the same relation (8.27) is valid with the obvious changes of the index, while if the process ($i \mapsto j$) is an ionization process it is

$$\sigma_{ij} (\epsilon) = \frac{\pi e^4}{\epsilon}\left(\frac{1}{\epsilon_{ji}} - \frac{1}{\epsilon}\right) \qquad (8.28)$$

whereas ϵ_{ji} is now the ionization energy from the level i.

If the colliding particles are neutral atoms the cross-section in the ionization process is much less known: from the experimental results obtained by Hayden and Amme [12] in the case of Ar, by Hayden and Utterbach [13] in the case of He it turns out that, at least in the limited interval of values of the ratio ϵ/ϵ_j. (ϵ_j ionization energy from the ground state) considered in those experiments, the experimental results can be expressed by the formula

$$(\sigma_j)_{at.} = 1.3\times 10^{-3} \frac{\pi e^4}{\epsilon^2}\left(\frac{\epsilon_{jH}}{\epsilon_j}\right)^2\left(\frac{\epsilon}{\epsilon_j} - 1\right)^2 \text{ cm}^2 \qquad (8.29)$$

being ϵ_{jH} the ionization energy of the H atom (from the ground state). In the case that the collisions lead to an excitation of the atom, not only there are no theoretical results, but also experimental results fail. As working hypothesis, and by analogy with the relations giving $(\sigma_{\kappa i})_{el.}$ and $(\sigma_{ij})_{el}$, one can assume

$$(\sigma_{ki})_{at.} = 1.3\times10^{-3}\,\frac{\pi e^4}{\epsilon_j^2}\left(\frac{\epsilon_{jH}}{\epsilon_j}\right)^2\left(\frac{\epsilon}{\epsilon_{ik}} - 1\right)^3 3\,f_{ki} \tag{8.30}$$

Finally, by means of the quantities above given it is possible to determine the radiative energy source for the electrons, that appears in the eq. (4.13). Reference is made to the processes represented by the notation (8.7): it appears that the electrons gain energy in the photoionization, and loose energy in the radiative recombination. In these processes one can find from the eq. (8.13) that the net radiative energy which is absorbed per unit time, per unit volume is given by rel.

$$\sum_i N_i \int_{\nu_{ji}}^{\infty} \sigma_{ij}\left[1-\frac{N_j^2}{N_i}\frac{\bar N_i}{\bar N_j^2}\exp\left(-h\nu/\kappa T_e\right)\right]\left[J_\nu - \frac{1-\exp(h\nu/\kappa T_e)}{\frac{\bar N_j^2}{\bar N_i}\frac{N_i}{N_j^2} - \exp(-h\nu/\kappa T_e)}B_\nu(T_e)\right]d\nu$$

being the summation extended over all the energy-levels (i); since the energy transferred to the atoms per unit volume to ionize them is expressed by $\sum_i (M_r^{(i)})_{ion.}\epsilon_{ji}$, where $(M_r^{(i)})_{ion.}$ is given by the rel. (8.15), one finds that the excess of radiative energy, which is absorbed, over that transferred to the atoms, and therefore the energy which is supplied to the free electrons is

$$Q_r^{(e)} = \sum_i \left\{N_i \int_{\nu_{ji}}^{\infty} \sigma_{ij}\left[1-\frac{N_j^2}{N_i}\frac{\bar N_i}{\bar N_j^2}\exp(-h\nu/\kappa T_e)\right]\cdot\right.$$

$$\left.\cdot\left[J_\nu - \frac{1-\exp(-h\nu/\kappa T_e)}{\frac{\bar N_j^2}{\bar N_i}\frac{N_i}{N_j^2} - \exp(-h\nu/\kappa T_e)}B_\nu(T_e)\right]d\nu\right\} - \tag{8.31}$$

$$-\sum_i (M_r^{(i)})_{ion}\,\epsilon_{ji}$$

For completeness, it is opportune to give also the collisional energy source for the electrons. This source is expressed by two parts: the one corresponding to the elastic collisions between electrons and heavier particles (atoms and ions), namely to the collisions that do not produce a change of the internal energy of these particles; the other corresponding to the anelastic collisions.

The elastic part $(Q_c^{(e)})_{el}$ is

$$(Q^{(e)})_{el.} = N_e \frac{m_e}{m_a} (\tau_{ej}^{-1} + \Sigma_i \tau_{e\alpha i}^{-1}) \kappa (T - T_e) \qquad (8.32)$$

whereas τ_{ej}^{-1} and $\tau_{e\alpha i}^{-1}$ are respectively the reciprocals of the characteristic times of collisions between electrons and ions and between electrons and atoms (in the state i), so defined

$$\tau_{e\alpha i}^{-1} = N_{\alpha i} \left(\frac{8\kappa T_e}{\pi m_e}\right)^{1/2} \sigma_{ea} \quad ; \quad \tau_{ej}^{-1} = N_j \left(\frac{8\kappa T_e}{\pi m_e}\right)^{1/2} \sigma_{ej} \qquad (8.33)$$

being

$$\sigma_{ea} = \begin{cases} [-0.35+0.775(10^{-4})\,T_e]\,10^{-16}\,\mathrm{cm}^2 & T_e > 10^4\,{}^\circ\mathrm{K} \\ [0.39-0.551(10^{-4})\,T_e + 0.595(10^{-8})\,T_e^2]\;10^{-16}\,\mathrm{cm}^2 & \\ & T_e \leqslant 10^4\,{}^\circ\mathrm{K}. \end{cases}$$

$$\sigma_{ej} = \frac{\pi e^4}{2(\kappa T_e)^2} \log n \left(\frac{9\kappa^3 T_e^{\,3}}{4\pi N_e e^6}\right)^{1/2} \qquad (8.34)$$

As for anelastic part $(Q_c^{(e)})_{in}$, one has to observe that in the collisional neutralization and diseccitation the energy gained by the electrons per unit time and unit volume is proportional respectively to the ionization energy and to the difference of the energies corresponding to the levels between which the transition occurs; on the other hand, to the same quantities it is also proportional the energy withdrawn from the electrons in the collisional ionization and excitation processes, when the colliding particles with the atoms are electrons.

Precisely it turns that

$$(Q_c^{(e)})_{in.} = - \Sigma_i \Sigma_k (M_c^{(e)})_{ki}\, \epsilon_{ik} - \Sigma_k (M_c^{(e)})_{kj}\, \epsilon_{kj}$$

where $(M^{(e)})_{ki}$ and $(M^{(e)})_{kj}$ are respectively the collisional source functions corresponding to the processes

$$A_k + e \rightleftarrows A_i + e \qquad (k < i) \;;\; A_k + e \rightleftarrows A^+ + e + e$$

9. Relative contributions of the radiative and collisional terms in the momentum, energy and mass transport and limit forms of the radiative contributions to the momentum and energy equations.

In order to determine the relative importance of the radiative contributions to the equations of transfer of momentum and energy it is opportune first of all to consider two limit cases, in which, these equations can be considerably simplified.

9.a. Rosseland approximation (or diffuse approximation).

Let be the case considered, in which the photon mean free path $l_\nu = (\kappa_\nu \rho)^{-1}$, according to the definition given in Sect. 7, is very small compared to the distance L from position P to the nearest region with appreciable difference in temperature, density or composition: such a region is called "optically thick", when the parameter (L/l_ν) is much greater than one, while the number (L/l_ν) is called "Bouguer number " of the radiation of frequency ν. Thus, for not very small values of s, it is $\tau_\nu \gg 1$ [see (7.2)]. Now when $\tau_\nu - (\tau_\nu)_\infty \gg 1$ the exponential factor in the second term at the right hand of the eq. (7.3) is very near to zero, so that it may be neglected; besides to the same degree of approximation, the lower integration limit in the first term can be replaced by $(-\infty)$. Finally, as a consequence of the assumption that the density ρ and the temperature T vary slightly over a length of the order of magnitude of one photon mean free-path, one can make a Taylor expansion of the function

$$F_\nu(\tau'_\nu) = \varphi(\tau'_\nu) B_\nu(\tau'_\nu) \quad \text{about} \quad F_\nu(\tau_\nu). \quad \text{One obtains}$$

$$F_\nu(\tau'_\nu) = F_\nu(\tau_\nu) + \left[\frac{dF_\nu(\tau'_\nu)}{d\tau'_\nu}\right]_{\tau'_\nu=\tau_\nu} (\tau'_\nu - \tau_\nu) + \frac{1}{2}\left[\frac{d^2 F_\nu}{d\tau'^2_\nu}\right]_{\tau'_\nu=\tau_\nu} (\tau'_\nu - \tau_\nu)^2 + \ldots \tag{9.1}$$

and replacing (9.1) in the eq. (7.3) allows to deduce

$$I_\nu(\tau_\nu) = \int_{-\infty}^{\tau_\nu} d\tau'_\nu \exp[-(\tau_\nu - \tau'_\nu)] \left\{ F_\nu(\tau_\nu) + (\tau'_\nu - \tau_\nu)\frac{dF_\nu}{d\tau_\nu} + \right.$$

$$\left. + \frac{1}{2}(\tau'_\nu - \tau_\nu)^2 \frac{d^2 F_\nu}{d\tau_\nu^2} + \ldots \right\} = F_\nu(\tau_\nu) - \frac{dF_\nu}{d\tau_\nu} + \frac{d^2 F_\nu}{d\tau_\nu^2} + \ldots \qquad (9.2)$$

But $d/d\tau_\nu = (1/\kappa_\nu|(d/ds)$ thus one obtains

$$I_\nu(s) = \varphi B_\nu - \frac{1}{\kappa_\nu}\frac{\partial}{\partial s}(\varphi B_\nu) + \frac{1}{\kappa_\nu}\frac{\partial}{\partial s}\left(\frac{1}{\kappa_\nu}\frac{\partial}{\partial s}(\varphi B_\nu)\right) + \ldots \qquad (9.3)$$

Relation (9.3) gives immediately

$$u^{(r)} = \frac{1}{c}\int_{4\pi} I_\nu \, d\Omega \cong \frac{4\pi}{c}\varphi B_\nu$$

$$\vec{F}_\nu^{(r)} = -\frac{1}{\kappa_\nu}\int_{4\pi} [\vec{s}\cdot \mathrm{grad}(\varphi B_\nu)]\vec{s}\, d\Omega = -\frac{1}{\kappa_\nu}\frac{4\pi}{3}\mathrm{grad}(\varphi B_\nu) \qquad (9.4)$$

$$p_{\nu_{xx}}^{(r)} = p_{\nu_{yy}}^{(r)} = p_{\nu_{zz}}^{(r)} \cong \frac{4\pi}{3c}\varphi B_\nu \; ; \; p_{\nu_{xy}}^{(r)} = p_{\nu_{xz}}^{(r)} = p_{\nu_{yz}}^{(r)} = 0$$

In the case that conditions of local thermodynamic equilibrium can be assumed, one can take $\varphi = 1$ and the rels. (9.4) become

$$u_\nu^{(r)} \cong \frac{4\pi}{c} B_\nu \; ; \; \vec{F}_\nu^{(r)} \cong \frac{-4\pi}{3\kappa_\nu}\frac{dB_\nu}{dT_e}\,\mathrm{grad}\, T_e \; ; \; p_\nu^{(r)} \cong \frac{4\pi}{3c} B_\nu \qquad (9.4')$$

These expressions correspond to the so called "Rosseland approximation", or "diffuse approximation".

Introducing the Rosseland mean-free path defined as

$$\ell_R = \frac{\int_0^\infty \ell_\nu \left(\frac{dB_\nu}{dT_e}\right) d\nu}{\int_0^\infty \frac{dB_\nu}{dT_e}\, d\nu} \qquad (9.5)$$

gives

$$E^{(r)} = \frac{1}{\rho} u^{(r)} = \frac{1}{\rho} a T^4 \quad \text{being} \quad a = \frac{8\pi^5 k^4}{15 c^3 h^3} = 76\times 10^{-15}\ \mathrm{erg\ cm^{-3}\ degree^{-4}}$$

$$\vec{F} = -\frac{4\pi}{3}\ell_R \frac{dB}{dT_e} \text{ grad } T_e = -\frac{4}{3} ac\, \ell_R T_e^3 \text{ grad } T_e = D^{(r)} \text{grad } u^{(r)}$$

(9.6)

writing

$$B = \int_0^\infty B_\nu\, d\nu = \frac{2k^4 T_e^4 \pi^4}{15c^2 h^3} \quad ;$$

(9.7) $D^{(r)} = c1_R/3$: the quantity $D^{(r)}$ is called the photon diffusion coefficient.

From the expression of $\vec{F}$ above given it turns out that the radiative contribution to the energy transfer takes the same form as that corresponding to the transfer by thermal conductivity, so that one can say that the radiation effect corresponds to increase the coefficient of thermal conductivity from the value Λ_t to the value $\Lambda_t + 4D^{(r)} a\, T_e^3$,or, in other words, the transfer of radiant energy from particle is formally analogous to ordinary heat conduction. It is clear therefore that, in the Rosseland approximation a region of the gas subjected to localized intensive heating may not suffer any severe loss of energy through radiation, the reason being that somewhat thicker region forms, enveloping the first region, such that the radiating energy is trapped.

Besides it appears that the relative contributions of the radiative energy and thermal energy transport are defined by the ratio

$$\frac{\frac{\ell_R B}{L}}{\Lambda_t \frac{T}{L}} \equiv O\left(\frac{\ell_R B}{L} \frac{L}{\mu c_p T} P_r\right) \equiv O\left(\frac{\ell_R B}{L} \frac{1}{\rho HU} P_r R_e\right) \equiv O\left(\frac{\ell_R B}{L} \frac{1}{\rho U^3} P_r R_e\right)$$

being P_r the Prandtl number of the mixture; R_e the Reynolds number referred to the velocity V and to the characteristic length L; μ the viscosity of the mixture, c_p its specific heat at constant pressure; H the enthalphy per unit mass. In writing the last relation it has been assumed that the enthalpy per unit mass is of the same order of the kinetic energy per unit mass.

Now, the ratio $\rho U^3/(1_R B/L)$, which is proportional to the ratio of the kinetic energy flux to the raditive energy flux (in the Rosseland approximation) is called the "Boltzmann number" of the field, and therefore one can conclude that the relative contribution of the radiative energy transfer in the energy equation is measured by the ratio of the product of the Prandtl and Reynolds numbers to the Boltzmann number.

Analogously one deduces that the order of magnitude of the ratio of the radiative pressure to the thermal pressure is given by

$$\frac{p^{(r)}}{p^{(t)}} \equiv 0\left(\frac{B}{ckN_o T}\right) \equiv 0\left(\frac{B}{\rho HU}\frac{U}{c}\right) \equiv 0\left(\frac{U}{c}\frac{B}{\rho U^3}\right)$$

if N_o is the number of the material particles of any species per unit volume. But, denoting with B_o the Boltzmann number it is $B/(\rho U^3) = (L/l_R)(1/B_o)$ and therefore

$$\frac{p^{(r)}}{p^{(t)}} \equiv 0\left(\frac{LU}{\ell_R c}\frac{1}{B_o}\right) \equiv 0\left(\frac{U}{c}\frac{B_{ou}}{B_o}\right)$$

if B_{ou} is the Bouguer number (referred to the Rosseland mean free path).

Now, it is $U/c \ll 1$ usually; therefore $p^{(r)}/p^{(t)} \ll 1$, generally; but, since in the Rosseland approximation $B_{ou} \ll 1$, it may occur that $p^{(r)}$ is not negligible with respect to $p^{(t)}$ when the Boltzmann number is sufficiently low.

More precisely, one obtains from the third of the rels. (9.4')

$$\frac{p^{(r)}}{p^{(t)}} \equiv \frac{aT^4}{3N_o kT} \equiv 17.2\,\frac{T^3}{N_o}$$

Thus, only for very high temperatures, as those involved in stellar structures ($T \geqslant 10^7$°K), or in space flight (N_o relatively small), the radiation pressure, and the radiative energy $E^{(r)}$, must be retained in the momentum or in the energy equations.

9.b. Planck approximation (or emission approximation)

Let us consider the opposite case to that corresponding to the Rosseland approximation, in which is satisfied the condition

$$\int_{s_o}^{s} \kappa_\nu \, ds \ll 1 \qquad (9.8)$$

and the medium is limited by a surface, from which the emission is negligible: if L is again a suitable length of the field, the first condition corresponds to assume that the Bouguer number (B_{ou}) is now very small, and combined with the second condition concerning the emission from the limiting surface leads to write

$$I_\nu(s) = \int_{s_0}^{s} \kappa_\nu \varphi\, B_\nu(s',T'_e)\,ds' \tag{9.9}$$

since $\exp[-(\tau_\nu-\tau'_\nu)]\cong 1$: the medium is called "optically thin". Moreover, if one assumes that the temperature T_e at s is not very much different from the temperature for much of the path between s_0 and s, it may be immediately deduced from rel. (9.9) that

$$I_\nu(s,T_e) \ll B_\nu(s,T_e) \tag{9.10}$$

The radiative energy transfer equation then becomes

$$\vec{s}\cdot \text{grad}\; I_\nu = \kappa_\nu \varphi\, B_\nu \tag{9.11}$$

from which it appears that the medium emits but not absorbs. It is easy to see from eq. (9.11) that the contribution by the radiation to momentum transport is null: in effect, multiplying both sides of the eq. (9.11) by $\vec{s}$ gives

$$\vec{s}\,(\vec{s}\cdot\text{grad}\; I_\nu) = \Sigma_k \frac{\partial}{\partial x_k}(I_\nu \ell_1 \ell_k)\,\vec{i}_1 + \Sigma_k \frac{\partial}{\partial x_k}(I_\nu \ell_2 \ell_k)\,\vec{i}_2 + $$

$$+ \Sigma_k \frac{\partial}{\partial x_k}(I_\nu \ell_3 \ell_k)\,\vec{i}_3 = \vec{s}\,\kappa_\nu\, \varphi\, B_\nu\,,$$

being l_j the component of $\vec{s}$ along the x_j axis and $\vec{i}_j$ the unit vector in the direction of the same axis. Multiplication by $d\Omega$ and integration over all directions allow to deduce

$$\vec{i}_1 \Sigma_k \frac{\partial}{\partial x_k}\int_{4\pi} \ell_1 \ell_k I_\nu\, d\Omega + \vec{i}_2 \Sigma_k \int_{4\pi} \ell_2 \ell_k I_\nu\, d\Omega + \vec{i}_3 \Sigma_k \frac{\partial}{\partial x_k}\int_{4\pi} \ell_3 \ell_k I_\nu\, d\Omega =$$

$$= \varphi \int_{4\pi} \vec{s}\,\kappa_\nu B_\nu\, d\Omega = 0$$

Thus

$$\Sigma_k \frac{\partial}{\partial x_k}\int_{4\pi} \ell_1 \ell_k\; I\, d\Omega = \Sigma_k \frac{\partial}{\partial x_k} p^{(r)}_{1k} = 0 \;;\; \Sigma_k \frac{\partial}{\partial x_k} p^{(r)}_{2k} = 0 \;;\; \Sigma_k \frac{\partial}{\partial x_k} p^{(r)}_{3k} = 0$$

and according to the eqs. (4.8), (8.1) the relations above given are just the radiative contributions to the momentum equation. There is, however, a radiative contribution to the energy transport: eq. (9.11) multiplied by $d\Omega$ and integrated over all solid angles and frequencies gives

$$\int_0^\infty d\nu \int_{4\pi} \vec{s}\cdot \text{grad } I_\nu \, d\Omega = \text{div} \int_0^\infty d\nu \int_{4\pi} \vec{s}\, I_\nu \, d\Omega = \text{div } \vec{F} = 4\pi\varphi \int_0^\infty \kappa_\nu B_\nu \, d\nu$$

The approximation corresponding to the expressions of the radiative stress tensor and energy flux above given is called "Planck approximation" or "emission approximation".

If κ_p is the "Planck mean opacity" defined by

$$\kappa_p = \frac{\int_0^\infty \kappa_\nu B_\nu \, d\nu}{\int_0^\infty B_\nu \, d\nu} = \frac{\int_0^\infty \kappa_\nu B_\nu \, d\nu}{\frac{\sigma T_e^4}{\pi}} \tag{9.12}$$

being σ the Stephan-Boltzmann constant

$$\left(\sigma = \frac{2\pi^5 k^4}{15 c^2 h^3} = 5.67\times 10^{-5} \text{ erg cm}^{-2} \sec^{-1} \text{degree}^{-4}\right)$$

one obtains

$$\text{div.}\vec{F} = 4\kappa_p \varphi \sigma T_e^4 \tag{9.13}$$

It is easy to see that even in this case it is

$$\frac{\text{div } \vec{F}}{\text{div } \vec{q}} \equiv 0 \left(\frac{P_r R_e}{B_o}\right) \tag{9.14}$$

Observe, on the other hand, that the radiative energy flux is in this case proportional to $\sigma T_e^4 L/1_p$, being $1_p = 1/\kappa_p$ the Planck photon mean free-path, or to (radiative energy flux by a black body) x B_{ou}.

9.c. General case.

Suppose that the absorption coefficient κ_ν is such that neither the condition corrispondent to the Rosseland approximation or the condition corrispondent to the Planck approximation is satisfied, while the emission from the surface that limits eventually the field is negligible.

Under such assumptions one finds from the eq. (7.3)

$$I_\nu = \int_{(\tau_\nu)_0}^{\tau_\nu} \varphi B_\nu \exp[-(\tau_\nu - \tau'_\nu)]\, d\tau'_\nu = (\overline{\varphi B_\nu}) \int_{\tau_{\nu_0}}^{\tau_\nu} \exp[-(\tau_\nu - \tau'_\nu)]\, d\tau'_\nu = \tag{9.15}$$
$$= (\overline{\varphi B_\nu})[1 - \exp(-\tau_{\nu_0})] \cong \overline{\varphi B_\nu}$$

where $\overline{\varphi B_\nu}$ denotes an appropriate average value of (φB_ν) along the ray to which the integration's path is extended, while it has been assumed $\tau_{\nu 0}$ sufficiently higher than the unity in order that $\exp(-\tau_{\nu 0})$ can be neglected. Thus one may deduce that in the more general case now considered the order of magnitude of I_ν in any limited domain of the field is of the order of magnitude of B_ν evaluated for a typical temperature T_m of the flow in that domain. So, from the expression of the radiative energy flux vector (8.6) one obtains that the component of this vector along the x axis is

$$F_{kx} = \int_{4\pi} I_\nu \cos\vartheta\, d\Omega = B_\nu(T_m) \int_{4\pi} \frac{I_\nu}{B_\nu} \cos\vartheta\, d\Omega = a_\nu B_\nu(T_m) \tag{9.16}$$

whereas a_ν is a coefficient depending on the frequency ν, that is of the order of unity: it is quite obvious that the x axis can always be chosen in such way that the component $F_{\nu x}$, which gives the radiative energy flux in the unit frequency interval, through the unit area perpendicular to the x axis, is representative of the order of magnitude of F_ν.

On the other hand it is

$$F_x = \int_0^\infty F_{\nu x}\, d\nu = \nu_j \int_0^\infty F_{\nu x}\, d\nu/\nu_j = \nu_j B_{\nu_j}(T_m) \int_0^\infty a_\nu \frac{B_\nu(T_m)}{B_{\nu_j}(T_m)}\, d\frac{\nu}{\nu_j} = \tag{9.17}$$
$$= \nu_j B_{\nu_j}(T_m) F_x^*$$

where ν_j is a typical frequency, such that $\int_0^\infty a_\nu [B_\nu(T_m)/B_{\nu_j}(T_m)]\, d\nu/\nu_j$ is of the order of unity.

Analogously, the order of magnitude of the thermal energy flux vector can be assumed to be equal to that of the component q_x.

$$q_x = -\Lambda_t \frac{\partial T}{\partial x} = (\Lambda_t)_m \frac{T_m}{L} = c_p \rho_m U_m T_m \frac{1}{P_r} \frac{1}{R_e} \tag{9.18}$$

where L is the characteristic length defined in (9.a); $(\Lambda_t)_m$ the value of the coefficient of the thermal conductivity at the characteristic temperature T_m; ρ_m and U_m are respectively characteristic density and characteristic velocity, while the Prandtl and Reynolds numbers P_r and R_e are referred to L , and to the values of the physical quantities corresponding to the temperature T_m. Thus it turns out that the relative contribution of the radiative and thermal terms in the energy equation is

$$\frac{F_x}{q_x} = 0\left(\frac{\nu_j B_{\nu j}}{\rho H_m U_m} P_r R_e\right) = 0\left(\frac{P_r R_e}{B_o}\right) \tag{9.19}$$

defining now the Boltzmann number B_o by means of the relation

$$B_o = \frac{\rho_m U_m^3}{\nu_j B_{\nu_j}}$$

and this result is analogous to that obtained in the Rosseland approximation. But, what is interesting in any fluid dynamics problem in order to characterize the influence of the radiation field on the flow of energy in the fluid is the ratio of $\partial F_x/\partial x$ to the correspondent rate of change of the thermal energy flux, or of the kinetic energy of the fluid.

Now, taking by analogy with the rel. (7.2)

$$dx = \kappa_{\nu_j} d\xi \tag{9.20}$$

one can write

$$\frac{\partial F_x}{\partial x} = \nu_j B_{\nu_j} \kappa_{\nu_j} \frac{\partial F^*}{\partial \xi}$$

and since

$$\frac{\partial q_x}{\partial x} = \frac{\partial}{\partial x}\left(\Lambda_t \frac{\partial T}{\partial x}\right) = 0\left((\Lambda_t)_m \frac{T_m}{L^2}\right)$$

it turns out that

$$\frac{\dfrac{\partial F_x}{\partial x}}{\dfrac{\partial q_x}{\partial x}} = 0\left(\frac{\nu_j B_{\nu j}}{\rho_m U_m H_m} P_r R_e \kappa_{\nu_j} L\right) = 0\left(\frac{P_r R_e}{B_o} \frac{L}{l_{\nu_j}}\right) \tag{9.21}$$

being $l_{\nu_j} = (\chi_{\nu_j})^{-1}$ the photon mean-free path corresponding to the frequency ν_j, so the ratio L/l_{ν_j} is the Bouguer number.

Thus it appears that the influence of the radiation field on the flow is depending on the ratio of the two parameters, the Bouguer and Boltzmann numbers. Consider now the radiative stress-tensor and assume as representative term of its order of magnitude the radiative pressure $p^{(r)}$: one can write

$$p^{(r)} = \frac{1}{3c}\int_{4\pi} I\, d\Omega = \frac{\nu_j B_{\nu j}(T_m)}{3c}\int_{4\pi} d\Omega \int_0^{\infty} \frac{I_\nu}{B_{\nu_j}}\, d(\nu/\nu_j) = \frac{\nu_j}{3c} B_{\nu_j}(T_m)\, P^{(r)} \tag{9.22}$$

so that the ratio of the radiative pressure to the thermal pressure has an order of magnitude given by

$$\frac{p^{(r)}}{p^{(t)}} = 0\left(\frac{\nu_j B_{\nu j}}{ck[(N_a + N_e)T_m + N_e T_{em}]}\right) = 0\left(\frac{\nu_j B_{\nu j}(T_m)}{\rho_m U_m U_m}\,\frac{U_m}{c}\right) = 0\left(\frac{1}{B_o}\,\frac{U_m}{c}\right) \tag{9.23}$$

Thus the relative importance of the radiative pressure and thermal pression appears to be measured by the product of the reciprocal of the Boltzmann number and the ratio U_m/c. However, what it is important in order to determine the influence of the radiation on the force exerted over a body in the flow field is not the pressure for it self, but its derivative, and in complete analogy with the relation (9.21) one finds

$$\frac{\dfrac{\partial p^{(r)}}{\partial x}}{\dfrac{\partial p^{(t)}}{\partial x}} = 0\left(\frac{B_{ou}}{B_o}\,\frac{U_m}{c}\right) \tag{9.24}$$

Since $U_m/c \ll 1$ generally, it appears that the force effect of the radiation can be generally neglected, except in the case $B_{ou}/B_o \gg 1$; in any case, by comparison of the rel. (9.24) with the rel. (9.21) one deduces that the influence of the radiation on the energy transport is generally much greater than the influence on the momentum transport.

In order to see the relative importance of the radiative and collisional source functions in the equation of mass transport it is opportune to consider simple cases as those of the ionization and excitation to the first level from the ground state.

In the case of ionization from the ground state, neglecting in the rel.

(8.15) the terms of the order of magnitude as $(-h\nu/kT)$, and taking $\sigma_{ij}=\sigma_{1j}$ by the Kramer's formula (6.32), one finds

$$(M_r^{(1)})_{ion.} = -4\pi N_1 \frac{32\pi^2 e^6}{3\sqrt{3}ch^3} \frac{\mathcal{R}_y}{\nu_j^3} B_{\nu_j} \int_1^\infty \left(\frac{\nu_j}{h\nu}\right) \frac{\nu_j^3}{\nu^3} \left[\frac{J_\nu}{B_{\nu_j}} - \frac{N_j^2}{N_1}\frac{\bar{N}_1}{\bar{N}_j^2}\frac{B_\nu}{B_{\nu_j}}\right] d\frac{\nu}{\nu_j} =$$

$$= -8\pi N_1 \frac{32\pi^2 e^6}{3\sqrt{3}c^3h^3} \mathcal{R}_y \exp(-T_j/T_e) \int_1^\infty \left(\frac{\nu_j}{\nu}\right)^4 \left[\frac{J_\nu}{B_{\nu_j}} - \frac{N_j^2}{N_1}\frac{\bar{N}_1}{\bar{N}_j^2}\frac{B_\nu}{B_{\nu_j}}\right] d\frac{\nu}{\nu_j} =$$

$$= -\frac{N_1}{\tau_{jr}} \int_1^\infty \left(\frac{\nu_j}{\nu}\right)^4 \left[\frac{J_\nu}{B_{\nu_j}} - \frac{N_j^2}{N_1}\frac{\bar{N}_1}{\bar{N}_j^2}\frac{B_\nu}{B_{\nu_j}}\right] d\frac{\nu}{\nu_j} \tag{9.25}$$

where $T_j = h\nu_j/k$ is the ionization temperature, and

$$(\tau_{jr})^{-1} = \frac{256}{3\sqrt{3}} \left(\frac{\pi e^2}{ch}\right)^3 \mathcal{R}_y \exp(-T_j/T_e) \tag{9.26}$$

On the other hand from the rels. (8.17), (8.26') one finds

$$(M_c^{(1)})_{ion.} = -N_1 \chi_{1j}\left(1 - \frac{N_j}{N_1}\frac{\chi_{j1}}{\chi_{1j}}\right) = -N_1\chi_{1j}\left(1 - \frac{N_j^2}{N_1}\frac{\bar{N}_1}{\bar{N}_j^2}\right) \tag{9.27}$$

and using the formulae (8.21'), (8.28)

$$(M_c^{(1)})_{ion} \cong -N_1 N_j \frac{1}{kT_e} \frac{4\pi e^4}{(2\pi m_e T_e)^{1/2}} \left(\frac{T_e}{T_j}\right)^2 \exp\left(-\frac{T_j}{T_e}\right)\left(1 - \frac{N_j^2}{N_1}\frac{\bar{N}_1}{\bar{N}_j^2}\right) =$$

$$= -\frac{N_1}{\tau_{jc}}\left(1 - \frac{N_j^2}{N_1}\frac{\bar{N}_1}{\bar{N}_j^2}\right) \tag{9.28}$$

This relation has been obtained assuming $(T_j/T_e) \gg 1$; besides, it is

$$(\tau_{jc})^{-1} = N_j \frac{1}{kT_e} \frac{4\pi e^4}{(2\pi m_e kT_e)^{1/2}} \left(\frac{T_e}{T_j}\right)^2 \exp(-T_j/T_e) \tag{9.29}$$

Thus, it turns out that

$$(9.30)\qquad \frac{(M_j^{(1)})_r}{(M_j^{(1)})_c} = \frac{\tau_{jc}}{\tau_{jr}} \frac{\displaystyle\int_1^{\infty} \left(\frac{\nu_j}{\nu}\right)^4 \left(\frac{J_\nu}{B_{\nu_j}} - \frac{N_j^2}{N_1}\frac{\bar{N}_1}{\bar{N}_j^2}\right) d\,\frac{\nu}{\nu_j}}{1 - \dfrac{N_j^2}{N_1}\dfrac{\bar{N}_1}{\bar{N}_j^2}}$$

Under the assumption that one can take, as for the evaluation of the order of magnitude of the integral term at the right side of the eq. (9.30), $J_\nu \cong B_\nu(T_{er})$ if T_{er} is an opportune temperature of the electrons in the region in which radiation propagates, it appears that the order of magnitude of the ratio $(M_j^{(1)})_r / (M_j^{(1)})_c$ is

$$(9.31)\qquad \frac{(M_j^{(1)})_r}{(M_j^{(1)})_c} = 0\left[\frac{\tau_{jc}}{\tau_{jr}} \frac{\left(1 - \dfrac{N_j^2}{N_1}\dfrac{\bar{N}_1}{\bar{N}_j^2}\right)\displaystyle\int_1^{\infty}\left(\frac{\nu_j}{\nu}\right)^4 \frac{B_\nu(T_{er})}{B_{\nu_j}(T_{er})}\, d\,\nu/\nu_j}{\left(1 - \dfrac{N_j^2}{N_1}\dfrac{\bar{N}_1}{\bar{N}_j^2}\right)}\right] =$$

$$= 0\left[\frac{\tau_{jc}}{\tau_{jr}}\left(\frac{T_{er}}{T_j}\right)\exp\left[-\frac{T_j}{T_e}\left(\frac{T_e}{T_{er}} - 1\right)\right]\right] \qquad \left(\text{assuming } \frac{T_{er}}{T_j} \ll 1\right)$$

Suppose that there exists a domain in which one can assume $T_{er} \cong T_e$: then, one obtains by the substitution of the expressions (9.26), (9.29) for τ_{jr} and τ_{jc} in the eq. (9.31) that the radiative source function is negligible in that domain with respect to the collisional source function when the following condition is satisfied

$$(9.32)\qquad N_j \gg \frac{256\pi^{3/2}}{12\sqrt{3}}\, \mathcal{R}_y^3 \left(\frac{T_j}{T_{jH}}\right)^{3/2} \left(\frac{T_j}{T_e}\right)^{-1/2}$$

being T_{jH} the ionization temperature of the hydrogen atom: thus f.i., taking $T_j = T_{jH}$; $T_j/T_e = 10$, the number of density of the electrons, in order that $(M_j^{(1)})_r$ be negligible with respect to $(M_j^{(1)})_c$, it has to be

$$N_j > 2.85\times10^{16}\ \text{cm}^{-3} \quad (\text{f.i.}\ \ N_j = N_e = 2.85\times10^{17}\ \text{cm}^{-3})$$

In this condition the flow is called to be "collision dominated": defining "degree of ionization" α_j the ratio $\alpha_j = N_j/(N_1 + N_j)$, and taking $No = N_1 + Nj = 26.87 \times 10^{18}\,\text{cm}^{-3}$, equal to the value of the number of atoms per cm^3 of a monoatomic gas at normal pressure and temperature, the value of N_j above indicated corresponds to a degree of ionization $\alpha_j = 2.85\times10^{17}/(26.87\times10^{18}) \cong 10^{-2}$.

Consider now the process of excitation from the ground state; the radiative source function is now (rel. 8.10').

$$(M_r)_{1,2} = -N_1 B_{12}\left(1 - \frac{N_2}{N_1}\frac{g_1}{g_2}\right)\left[C_{\nu_{21}}\int_0^\infty \Phi_\nu J_\nu \, d\nu/\nu_{21} - \frac{N_2}{N_1}\frac{\bar{N}_1}{\bar{N}_2} B_{\nu_{21}}\right] \tag{9.33}$$

neglecting once again the terms of order of magnitude as $\exp(-h\nu_{21}/kT)$. Defining a degree of excitation α_e by means of the rel.

$$\alpha_e = \frac{N_2}{N_1 + N_2} = \frac{N_2}{N_0} \tag{9.34}$$

one can write

$$(M_r)_{1,2} = -N_0\left[1 - \alpha_e\left(1 + \frac{g_1}{g_2}\right)\right]\frac{4\pi^2 e^2}{m_e c} f_{12}\frac{B_{\nu_{21}}}{h\nu_{21}}\left(C_{\nu_{21}}\int_0^\infty \Phi_\nu \frac{J_\nu}{B_{\nu_{21}}} d\nu/\nu_{21} - \right.$$

$$\left. - \frac{\alpha_e}{1-\alpha_e}\frac{1-\bar{\alpha}_e}{\bar{\alpha}_e}\right) = -\frac{N_0\left[1-\alpha_e(1+g_1/g_2)\right]}{\tau_{er}}\left(C_{\nu_{21}}\int_0^\infty \Phi_\nu \frac{J_\nu}{B_{\nu_{21}}} d\nu/\nu_{21} - \frac{\alpha_e}{1-\alpha_e}\frac{1-\bar{\alpha}_e}{\bar{\alpha}_e}\right) \tag{9.35}$$

taking into account the rel. (6.14) and being

$$(\tau_{er})^{-1} = \frac{4\pi^2 e^2}{m_e c} f_{12}\frac{B_{\nu_{21}}(T_e)}{h\nu_{21}} . \tag{9.36}$$

On the other hand the collisional source function for the excitation process under consideration is

$$(M_c)_{1,2} = -N_1 \chi_{12}\left(1 - \frac{N_2}{N_1}\frac{\bar{N}_1}{\bar{N}_2}\right) = -N_0(1-\alpha_e)\chi_{12}\left(1 - \frac{\alpha_e}{1-\alpha_e}\frac{1-\bar{\alpha}_e}{\bar{\alpha}_e}\right) \tag{9.37}$$

Thus, in the more important case, in which the colliding particles are atoms and electrons, from the rels. (8.21), (8.21') one finds

$$(M_c)_{1,2} = -N_1 N_e (1-\alpha_e)\frac{12\pi e^4 f_{12}}{kT_e(2\pi m_e T_e)^{1/2}}\left(\frac{T_e}{T_*}\right)^2$$

(9.38) $$\exp\left(-\frac{T_*}{T_e}\right) \times \left(1 - \frac{\alpha_e}{1-\alpha_e}\,\frac{1-\bar{\alpha}_e}{\bar{\alpha}_e}\right)$$

where T_* is the excitation temperature($T_* = h\nu_{21}/k$) , and terms of the order of magnitude as $\exp(-T_*/T)$ have been neglected with respect to unity.

One can write

(9.38') $$(M_c)_{1,2} = -\frac{N_o(1-\alpha_e)}{\tau_{ec}}\left(1 - \frac{\alpha_e}{1-\alpha_e}\,\frac{1-\bar{\alpha}_e}{\bar{\alpha}_e}\right)$$

being

(9.39) $$(\tau_{ec})^{-1} = N_e \frac{12\pi e^4 f_{12}}{kT_e(2\pi m_e kT_e)^{1/2}}\left(\frac{T_e}{T_*}\right)^2 \exp(-T_*/T_e)$$

The order of magnitude of the ratio $(M_r)_{1,2}/(M_c)_{1,2}$ turns out to be

(9.40) $$\frac{(M_r)_{1,2}}{(M_c)_{1,2}} = 0\left[\frac{\tau_{ec}}{\tau_{er}}\;\frac{C_{\nu_{21}}\displaystyle\int_0^\infty \Phi_\nu \frac{J_\nu}{B_{\nu_{21}}}\,d\nu/\nu_{21} - \frac{\alpha_e}{1-\alpha_e}\,\frac{1-\bar{\alpha}_e}{\bar{\alpha}_e}}{1 - \dfrac{\alpha_e}{1-\alpha_e}\,\dfrac{1-\bar{\alpha}_e}{\bar{\alpha}_e}}\right]$$

As for the evaluation of the order of magnitude, one can assume the Φ_ν function as a Dirac function $\delta(\nu-\nu_{21})$, and therefore $C_{\nu_{21}} \cong 1$; $\int_0^\infty \Phi_\nu J_\nu / B_{\nu_{21}}(T_e)\, d\nu/\nu_{21} \cong J_{\nu_{21}}/B_{\nu_{21}}$; so that

(9.41) $$\frac{(M_r)_{1,2}}{(M_c)_{1,2}} = 0\left[\frac{\tau_{ec}}{\tau_{er}}\;\frac{\left(\dfrac{J_{\nu_{21}}}{B_{\nu_{21}}}\right) - \dfrac{\alpha_e}{1-\alpha_e}\,\dfrac{1-\bar{\alpha}_e}{\bar{\alpha}_e}}{1 - \dfrac{\alpha_e}{1-\alpha_e}\,\dfrac{1-\bar{\alpha}_e}{\bar{\alpha}_e}}\right]$$

Under the same assumption made in the previous case, and with the same meaning of the symbol T_{er}, the eq. (9.41) becomes

$$\frac{(M_r)_{1,2}}{(M_c)_{1,2}} = 0\left[\frac{\tau_{ec}}{\tau_{er}} \frac{\exp\left[-\left(T_*/T_e\right)\left(\frac{T_e}{T_{er}}-1\right)\right] - \frac{\alpha_e}{1-\alpha_e}\frac{1-\bar{\alpha}_e}{\bar{\alpha}_e}}{1-\frac{\alpha_e}{1-\alpha_e}\frac{1-\bar{\alpha}_e}{\bar{\alpha}_e}}\right] \quad (9.42)$$

From the expressions (9.36), (9.39) of the characteristic times τ_{er} and τ_{ec} one finds

$$\frac{\tau_{ec}}{\tau_{er}} = 8.5\,\frac{\mathcal{R}_y^3}{N_e}\left(\frac{T_*}{T_e}\right)^{1/2} \quad (9.43)$$

and therefore in order that the excitation process under consideration be collision dominated in a region in which one can assume $T_e = T_{er}$, the following condition has to be satisfied

$$8.5\,\frac{\mathcal{R}_y^3}{N_e}\left(\frac{T_*}{T_e}\right)^{1/2} \ll 1 \quad (9.44)$$

Assuming $T_*/T_e = 10$, one deduces $N_e \gg 0.35 \times 10^{17}$ cm^{-3}; taking $N_e = 0.35 \times 10^{18}$ cm^{-3} , since at normal temperature and pressure it is $N_0 = 2.687 \times 10^{19}$ cm^{-3} the correspondent degree of ionization turns out to be $\alpha_j = 0.013$.

10. Boundary conditions

10.a. Boundary condition at infinity.

The general equations of the flow of a radiating gas have been written, at least formally; but, in order to specify the solution of any given problem it is still necessary to give the appropriate boundary conditions: those pertaining to the physical quantities related to the mass particles are generally known from the mechanics of non radiating fluids, but in presence of a radiation field other conditions have to be written.

One case is very simple to consider: it is the one in which the straight line, or the plane at infinity belongs to the boundary, since in this case the correspondent condition is simply

$$J_\nu = B_\nu \quad (10.1)$$

which does not express any thing else than the system is in equilibrium at infinity. But, if portion of the boundary is formed by solid walls the correspondent conditions are related to the problem of the interaction of radiation with solid surfaces, which is a very complicated problem and subject of specialized studies: here only a gross description of the phenomenon will be given, which can be sufficient to provide simplified or approximated boundary conditions for the treatment of the influence of radiation in gasdynamic.

10.b. Boundary condition on a surface. Reflectivity, absorptivity, transparency.

Consider radiation that is incident upon a plane surface which limits a slab of material, and assume this surface Σ as coordinate plane (y, z) being the orientation of the x axis coincident with the outward normal to the surface itself.

This assumption is not too restrictive, because, with the ecception of singular points, any surface in the neighborhood of a given point can be replaced by its tangent plane in this point, so that in the general case one has to assume that the plane (y, z) here considered is the tangent plane in a certain point at the surface which limits the flow field.

The symbol I_ν^- with a superscript – denotes intensity corresponding to radiation that strikes the surface Σ, while the same symbol with a superscript +, (I_ν^+), indicates intensity of radiation which leaves Σ. It is assumed that I_ν^- and I_ν^+ may vary from point to point of the plane (y, z); furthermore, the attention is restricted to cases such that the frequency of radiation is not changed as a result of the interaction with Σ.

Now the intensity I_ν^- of radiation striking Σ at the origin can be written

$$I_\nu^-(0) = r'_{s\nu}(0)\, I_\nu^-(0) + \alpha'_{s\nu}(0)\, I_\nu^-(0) + \lambda'_{s\nu}(0)\, I_\nu^-(0) \tag{10.2}$$

where the subscript s indicates that the quantities are depending on the properties of the material of the slab; moreover $r'_{s\nu}\, I_\nu^-$ is the part of I_ν^- which is reflected by the surface, $\alpha'_{s\nu} I_\nu^-$ is the part absorbed by the medium S, and $\lambda'_{s\nu} I_\nu^-$ is the part transmitted through this medium. The coefficients $r'_{s\nu}$, $\alpha'_{s\nu}$, $\lambda'_{s\nu}$ are called the spectral reflectivity, absorptivity, and transparency of the material S, and in general they depend on the point being considered on the surface Σ of S.

It is obvious from eq. (10.2) that

$$r'_{s\nu} + \alpha'_{s\nu} + \lambda'_{s\nu} = 1. \tag{10.3}$$

If the transparency is null, the surface is called "opaque" and this is the only case that will be considered, so that for the surfaces Σ under consideration it is

$$r'_{\alpha s} + \alpha'_{s\nu} = 1 \tag{10.4}$$

Let be $\vec{s}$ the unit vector in the propagation direction of the radiation striking Σ ; θ the angle that $\vec{s}$ forms with the x axis, and φ the angle that the projection of $\vec{s}$ in the plane (y, z) forms with the y axis. The symbol $\vec{s}'$ is used to define the propagation direction of the radiation leaving the surface, and θ', φ' denote the corresponding angles; finally μ is used to denote $\mu = \cos\theta$, while $\mu' = \cos\theta$.

The probability for photons with initial direction $\vec{s}$ to be reflected by the surface Σ into the element of solid angle $d\Omega'$ centered about the direction $\vec{s}'$ will be denoted by the symbol $P_{s\nu}(\mu, \varphi; \mu', \varphi')\, d\Omega'$.

The number of photons with frequency in the range ν to $\nu + d\nu$ leaving per unit area of the surface at x = o, per unit time, with direction of propagation within the element of solid angle $d\Omega'$ about the unit vector $\vec{s}'$ is given by

$$\mu' (h\nu)^{-1} \; I^+_\nu (0,\mu',\varphi')\, d\Omega' d\nu' \quad ; \tag{10.5}$$

the number of photons with frequency in the same range, that strike the unit area at x=0 in the unit time with direction of propagation within the solid angled Ω about $\vec{s}$ is

$$\mu (h\nu)^{-1} \; I^-_\nu(0,\mu,\varphi)\, d\Omega\, d\nu \tag{10.5'}$$

The portion of these striking photons, which is reflected within the solid angle $d\Omega'$ is therefore given by

$$-\mu (h\nu)^{-1} I^-_\nu(0,\mu,\varphi)\, d\Omega\, d\nu \;\; P_{s\nu}(\mu,\varphi;\; \mu',\varphi')\, d\Omega' \tag{10.6}$$

and the total number of photons ν reflected within $d\Omega'$ results to be

$$-\int_{\Omega^-} \mu \;(h\nu)^{-1} \; I^-_\nu(\mu,\nu) \;\; P_{s\nu}(\mu,\varphi;\; \mu',\varphi' \;)\, d\nu \;\; d\Omega' d\Omega \tag{10.7}$$

If the surface were perfectly reflective the quantities (10.7) and (10.5) should be equal, so that in the case of perfect reflectivity it is

$$\mu' I^+_\nu (\mu',\varphi') = -\int_{\Omega^-} \mu \; I^-_\nu (\mu,\varphi) \;\; P_{s\nu} (\mu,\varphi;\; \mu',\varphi')\, d\Omega \tag{10.8}$$

being Ω^- the hemisphere of radius one, centered at 0 (the point of incidence). Let be

$$r''_{s\nu} (\mu,\varphi;\; \mu',\varphi') = \mu \, P_{s\nu} (\mu,\varphi;\; \mu',\varphi') \tag{10.9}$$

It has been proposed by R.V. Dunkle [14] to call $r''_{s\nu}$ the "biangular" spectral reflectance and it can be shown that a reciprocity relationship holds for the biangular reflectance, that is

$$r''_{s\nu}(\mu,\varphi;\mu',\varphi') = r''_{s\nu}(\mu',\varphi';\mu,\varphi).$$

Replacing the expression (10.9) in the rel. (10.8) one finds

(10.8') $$I^+_\nu(\mu',\varphi') = -\frac{1}{\mu'}\int_{\Omega^-} I^-_\nu(\mu,\varphi)\, r''_{s\nu}(\mu,\varphi;\mu',\varphi')\, d\Omega$$

in the case of perfect reflectivity: if the surface Σ is not a perfect reflector, one can write for the intensity of radiation which leaves the surface in the $\vec{s}'$ direction

(10.10) $$I^+_\nu(\mu',\varphi') = -\frac{1}{\mu'}\int_{\Omega^-} I^-_\nu(\mu,\varphi)\, r''_{s\nu}(\mu,\varphi;\mu',\varphi')\, d\Omega + S_\nu(\mu',\varphi')$$

where $S_\nu(\mu',\varphi')$ is the contribution to $I^+_\nu(\mu',\varphi')$ due to the emissivity by Σ. Now, the portion of the striking flux which is reflected by Σ is, according to (10.2)

(10.10') $$-\int_{\Omega^-} r'_{s\nu}(\mu,\varphi)\,\mu\, I^-_\nu(\mu,\varphi)\, d\Omega$$

while from (10.10) one deduces that the portion of the leaving flux which represents the contribution due to the reflection by the surface is

(10.11) $$-\int_{\Omega^+}\mu'\left[\frac{1}{\mu'}\int_{\Omega^-} I^-_\nu(\mu,\varphi)\, r''_{s\nu}(\mu,\varphi;\mu',\varphi')\, d\Omega\right] d\Omega' = -\int_{\Omega^-} I^-_\nu(\mu,\varphi)\, d\Omega\int_{\Omega^+} r''_{s\nu}(\mu,\varphi;\mu',\varphi')\, d\Omega'$$

where Ω^+ represents the emisphere corresponding to direction of propagation $\vec{s}'$. Since the quantities (10.10') and (10.11) represent the same flux, it must be

(10.12) $$\mu\, r'_{s\nu}(\mu,\varphi) = \int_{\Omega^+} r''_{s\nu}(\mu,\varphi;\mu',\varphi')\, d\Omega'$$

Let be

(10.13) $$r''_{s\nu}(\mu,\varphi;\mu',\varphi') = \mu\,\mu'\, r^*_{s\nu}(\mu,\varphi;\mu',\varphi')$$

where it is still

$$r^*_{s\nu}(\mu,\varphi;\mu',\varphi') = r^*_{s\nu}(\mu',\varphi';\mu,\varphi)$$

Rel. (10.12) becomes

(10.12') $$r'_{s\nu}(\mu,\varphi) = \int_0^1\int_0^{2\pi} \mu'\, r^*_{s\nu}(\mu,\varphi;\mu',\varphi')\, d\mu'\, d\varphi'$$

Two cases are particularly simple, and interesting: suppose that

$$r^{*}_{s\nu}\ (\mu,\varphi;\ \mu',\varphi') = \frac{1}{\mu'}\delta\ (\mu+\mu')\ \delta(\varphi-\varphi')\ b_\nu \tag{10.14}$$

being $\delta\,(x-a)$ the Dirac function: this function has the property that

$$\int_{a-\Delta x}^{a+\Delta x} f(x)\,\delta(x-a)\,dx = f(a)$$

for any Δx

the symbol b_ν denotes a constant. Thus the first term at the right-hand side of eq. (10.10) becomes

$$-\frac{1}{\mu'}\int_{\Omega^-} r''_{s\nu}\ (\mu,\varphi;\ \mu',\varphi')\ I^-_\nu(\mu,\varphi)\ d\Omega = \int_0^{-1}\int_0^{2\pi} b_\nu\ \frac{\mu}{\mu'}\ I^-_\nu(\mu,\varphi)\ \delta\ (\mu+\mu')\ \cdot$$

$$\cdot\ \delta\ (\varphi-\varphi')\ d\mu\ d\varphi = b_\nu\ I^-_\nu(-\mu',\varphi')$$

and therefore this case corresponding to the assumption (10.14) is that of the specular reflection.

Suppose now that

$$r^{*}_{s\nu}\ (\mu,\varphi;\ \mu',\varphi') = \text{const.} = a_\nu \tag{10.14'}$$

The eq. (10.12') gives

$$r'_{s\nu} = a_\nu\,\pi$$

and therefore $r^{*}_{s\nu} = r'_{s\nu}/\pi$. Thus the first term at the right-hand side of eq. (10.10) becomes

$$-\frac{1}{\mu'}\int_{\Omega^-} r''_{s\nu}\ (\mu,\varphi;\ \mu',\varphi')\ I^-_\nu(\mu,\varphi)\ d\Omega = \frac{1}{\pi}\int_0^{-1}\int_0^{2\pi} r'_{s\nu}\ \ I^-_\nu(\mu,\varphi)\,\mu\,d\mu\,d\varphi \tag{10.15}$$

and this relation says that the part of $I^+_\nu\ (\mu',\varphi')$ due to the reflection is independent on μ' and φ': therefore the assumption (10.14') corresponds to the case of perfectly diffuse reflection.

It is noteworthy to observe that the assumptions above indicated are examples of the more general case in which the probability function $r^{*}_{s\nu}$ may be expressed by means of a series of the type

$$r^*_{s\nu}\ (\mu,\varphi;\ \mu',\varphi') = \frac{b_\nu}{\mu'}\ \delta\ (\mu + \mu')\ \delta\ (\varphi - \varphi') + \sum_{\ell=0}^{\infty} a_{\ell\nu}\ P_\ell\ (\cos\ \Theta) \tag{10.16}$$

whereas the coefficients b_ν , $a_{\ell\nu}$ depend on the frequency solely; P_ℓ is the Legendre polynomial of degree ℓ; Θ is the angle that the reflected radiation forms with the incident radiation.

10.c. Einstein photoelectric effect. Determination of the source function S_ν .

As for the contribution $I^+_\nu\ (\mu', \varphi')$ due to the emission, only the case, in which the emitting medium is a metal, will be considered here: now it is well known that when a metal is irradiated by light, whose photons have an amount of energy in excess of a critical value, electrons are emitted: the critical frequency ν_c is defined by the relation

$$h\nu_c = e\Phi \tag{10.17}$$

where Φ is the photoelectric work function, while when $\nu > \nu_c$ is satisfied the Einstein's photoelectric equation (which is completely analogous to rel (3.31))

$$\frac{m_e v^2}{2} = h\nu - e\Phi \tag{10.18}$$

being v the maximum velocity of the emitted electron. In the invers process, when an electron striking on the surface is captured by an ion, a photon of frequency ν given by the eq. (10.18) is emitted.

Let be $\Psi_1^{(2)}$ the probability that a light quantum in a radiation field of unit specific intensity be captured on the surface Σ: assume that the number of photons absorbed per unit time, per unit area of the surface Σ, per unit solid angle, and per unit frequency interval may be expressed by means of the relation

$$N_a = (n_{es})^q\ \psi_1^{(2)}\ I^-_\nu \tag{10.19}$$

being n_{es} the number of free electrons per unit volume in the metal and q an appropriate constant, so that the corresponding absorbed energy is (10·19') $E_a = h\nu\ (n_{es})^q\ \Psi_1^{(2)} I^-_\nu$, while the corresponding number of electrons emitted by photoelectric effect is

$$2\pi (n_{es})^q \int_{\nu_c}^{\infty} d\nu \int_{\Omega^-} \psi_1^{(2)} I_\nu^- \, d\Omega = \int_{\nu_c}^{\infty} (J_\nu^-)^* \, d\nu$$

being $(J_\nu^-)^*$ defined by the rels. $(J_\nu^-)^* = \frac{1}{2\pi} \int_{\Omega^-} \psi_1^{(2)} I_\nu^- \, d\Omega$

Let be $\alpha_2^{(1)}$ the probability per second, per incident free electron in the energy range ϵ to $\epsilon + d\epsilon$ that an electron be captured on the surface Σ and a photon be given off, and assume that the number of photons emitted per second and per unit area within the unit solid angle, per unit frequency interval is equal to

$$\frac{1}{2\pi} (n_{es})^q N_e f(\epsilon) \alpha_2^{(1)} h \sqrt{\frac{2\epsilon}{m_e}}$$

being $f(\epsilon)$ the energy distribution function of the incident electrons, so that the corresponding emitted energy is

$$E_e = \frac{h\nu}{2\pi} (n_{es})^q N_e f(\epsilon) \alpha_2^{(1)} h \sqrt{\frac{2\epsilon}{m_e}} \qquad (10.19'')$$

while the corresponding number of electrons captured per unit area, per unit time is

$$(n_{es})^q N_e \int_0^{\infty} f(\epsilon) \left(\alpha_2^{(1)}\right)^* \sqrt{\frac{2\epsilon}{m_e}} \, d\epsilon$$

being N_e the number of density of the electrons in the gas stream and $(\alpha_2^{(1)})^* = 1/2\pi \int_{\Omega^+} \alpha_2^{(1)} d\Omega$. Consider not the system in condition of thermodynamic equilibrium: the emitted and absorbed energies are equal, and since, under this condition it is $I_\nu^- = B_\nu$ it turns out that

$$\frac{1}{2\pi} (\bar{n}_{es})^q \bar{N}_e f(\epsilon) \alpha_2^{(1)} h^2 \nu \left(\frac{2\epsilon}{m_e}\right)^{1/2} = (\bar{n}_{es})^q \psi_1^{(2)} B_\nu h\nu$$

Suppose that the energy distribution function $f(\epsilon)$ is the Maxwellian one (and critical considerations about this assumption will be made later on), so that

$$f(\epsilon) = \frac{2}{\pi^{1/2}} (kT_e)^{-3/2} (\epsilon)^{1/2} \exp\left(-\frac{\epsilon}{kT_e}\right) =$$

$$= \frac{2}{\pi^{1/2}} (kT_e)^{-3/2} (\epsilon)^{1/2} \frac{c^2}{2h\nu^3} B_\nu \exp\left(\frac{h\nu_c}{kT_e}\right)$$

since

$$\epsilon = h\nu - \epsilon_c; \quad \exp\left(-\frac{h\nu}{kT_e}\right) = \frac{c^2}{2h\nu^3} B_\nu (T_e)$$

Thus one obtains

$$\frac{1}{2\pi}\,\bar{N}_e\,\alpha_2^{(1)}\,h^2\nu\,\frac{2}{\pi^{1/2}}\,(kT_e)^{3/2}\,\epsilon^{1/2}\,\frac{c^2}{2h\nu^3}\exp\left(\frac{h\nu_c}{kT_e}\right)B_\nu(T_e)\left(\frac{2\epsilon}{m_e}\right)^{1/2} =$$

$$= \psi_1^{(2)}\,h\nu B_\nu(T_e)$$

Let be the exponential q in the expressions (10.19), (10.19') such that the probability factors $\Psi_1^{(2)},\alpha_2^{(1)}$ are not depending on the state of equilibrium or disequilibrium of the system; one finds

$$\text{(10.20)}\qquad \left.\begin{aligned} \bar{N}_e &= 2\,\frac{(\pi m_e k T_e)^{3/2}}{h^3}\exp\left(-\frac{h\nu_c}{kT_e}\right)\\ \psi_1^{(2)} &= \sqrt{2}\,\alpha_2^{(1)}\,\frac{(m_e c)^2}{(h\nu)^3}\left(\frac{\epsilon}{m_e}\right)\end{aligned}\right|$$

Now, according to the rel. (10.2) the same amount of the energy expressed by the rel. (10.19') is expressed by the second term at the right-hand side of the rel. (10.2), so that one can write

$$\text{(10.21)}\qquad \alpha'_{s\nu} = 1 - r'_{s\nu} = h\nu\,(n_{es})^q\,\psi_1^{(2)}$$

On the other hand it is apparent from what above indicated that the "source term" S_ν at the right-hand side of the eq. (10.2) is given by

$$S_\nu = \frac{h^2\nu}{2\pi}(n_{es})^q N_e f(\epsilon)\alpha_2^{(1)}\left(\frac{2\epsilon}{m_e}\right)^{1/2} = h\nu\,\frac{h^3\exp(h\nu_c/kT_e)}{2(\pi m_e kT_e)^{3/2}}(n_{es})^q N_e\,\psi_1^{(2)}\,B_\nu =$$

$$\text{(10.22)}\qquad = \alpha'_{s\nu}\,\frac{N_e}{\bar{N}_e}\,B_\nu = \beta'_{s\nu}\,B_\nu(T_w)$$

since at the surface Σ the temperatures of all particles have to be equal to the temperature T_w of the wall, and where the coefficient $\beta'_{s\nu}$ is called emissivity of the surface under consideration.

One can see immediately that $\alpha'_{s\nu} = \beta'_{s\nu}$ if the system is in thermodynamic equilibrium, and this relation expresses the Kirchoff's law.

Suppose that the wall is electrically insulated: then in stationary conditons the number of the electrons that escape from the surface Σ has to be equal to the number of the electrons which recombines with the ions on the surface. It has to be observed right now that only under the condition now specified the hypothesis that the electrons incident on Σ are in equilibrium as for the traslational degree of freedom may be accepted; on the contrary, if the wall is not insulated, but it is a part of an electrical circuit, the number of the emitted electrons can be much different from that of the incident electrons on the surface, and therefore near the surface the kinetic energy distribution function for those electrons can be completely out of equilibrium. At any way, under the assumption of electrical insulation of the wall one can write

$$2\pi \int_0^\infty (J_\nu^-)^* d\nu = N_e \int_0^\infty f(\epsilon)\left(\alpha_2^{(1)}\right)^* \sqrt{\frac{2\epsilon}{m_e}}\, d\epsilon \qquad (10.23)$$

Thus the eq. (10.23) defined the value of the quantity N_e on the wall and since the value of the same quantity but in condition of thermodynamic equilibrium is given by the rel. (10.20), the source term S_ν can be evaluated, and the eq. (10.10) combined with the rel. (10.12') gives the boundary condition concerning the specific intensity I_ν provided the biangular spectral reflectance $r''_{s\nu}$ is known.

As far as this subject is concerned, it has to be observed that a satisfactory theory of reflectivity and emissivity of surface for any material does not exist: a large number of theoretical and experimental informations exists for metals (see f.i. [15] , [16]). However, they are concerned not with the $r^*_{s\nu}$ factor, but with the spectral reflectivity $r'_{s\nu}$. The theoretical researches by Price [16] give the following dependence of the angular reflectivity $r'_{s\nu}$ on the frequency ν for normal incidence:

$$r'_{s\nu} = \frac{1}{16}\left(\frac{\nu_o}{\nu}\right)^4 , \text{ when } \nu > \nu_o , \text{ being } \nu_o = \left(\frac{n_{es}}{\pi m_e}\right)^{1/2} e \;; \qquad (10.24)$$

n_s the number of free electrons per unit volume (in the metal) (on the approximate basis of one free electron per atom ν_o is of the order of 10^{15} sec^{-1});

$$r'_{s\nu} = 1 - 2\frac{\nu_o}{\nu_r} \quad \text{when } \frac{\nu_o^2}{\nu_r} < \nu < \nu_o , \text{ being } \nu_r = 2\sigma \qquad (10.24')$$

where σ denoted the electrical conductivity for steady currents (ν_r is of the order of 10^{17} sec^{-1} for most metals);

$$r'_{s\nu} = 1 - 2\left(\frac{2\nu}{\nu_r}\right)^{1/2} \quad \text{when } \nu < \frac{\nu_o^2}{\nu_r} \qquad (10.24'')$$

(and therefore, taking for ν_r and ν_o the values above indicated, when $\nu < 10^{13}$).

Since, when ν is just equal to 10^{13} it is $2\nu/\nu_r = 2\times10^{-4}$ and therefore $r'_{s\nu} = 1-2\sqrt{2}\times 10^{-2} = 0{,}975$, while for $10^{13} \leqslant \nu \leqslant 10^{15}$ it is $r'_{s\nu} = 0{,}98$, it appears from the relations above given that a critical frequency exists for the reflectivity ν_0, such that below the frequency ν_0, $r'_{s\nu}$ differs little from one, while above ν_0, $r'_{s\nu}$ decreases very rapidly, almost discontinuously, as it is shown in the diagramms of fig. 6.

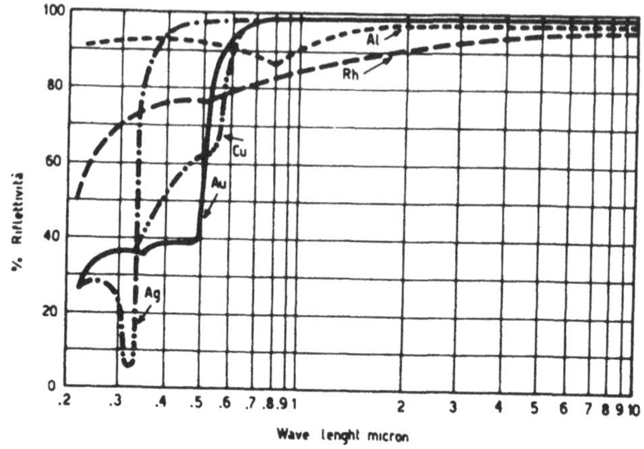

Fig.6

In addition, one may observe that the electrical conductivity σ can be expressed by means of the rel. (Drude's equation)

$$\sigma = \frac{n_{es}\, e^2}{m_e} \frac{\ell_o}{v_o}$$

where v_0 denotes Fermi's limiting velocity and ℓ_0 is the mean free path for electrons moving with that velocity.

It turns out that

$$\alpha'_{s\nu} = 1 - r'_{s\nu} = A_o \frac{\nu^{\frac{1}{2}}}{n_{es}^{\frac{1}{2}}} \quad , \quad \text{for } \nu < \frac{\nu_o^2}{\nu_r}$$

$$\alpha'_{s\nu} = A_1 \frac{1}{n_{es}^{\frac{1}{2}}} \quad , \quad \text{for } \frac{\nu_o^2}{\nu_r} < \nu < \nu_o$$

$$\alpha'_{s\nu} \cong 1 \quad , \quad \text{for } \nu > \nu_o$$

If one compares these expressions of $\alpha'_{s\nu}$ with that given by the rels. (10.21) one finds that in order the probability coefficients $\Psi_1^{(2)}$ and $\alpha_2^{(1)}$ are not depending on n_{es} the exponent q has to be

$$q = -1/2 \quad \text{for } \nu < \nu_0 \quad ; \quad q = 0 \quad \text{for } \nu > \nu_0$$

As far as the dependance on the direction $\vec{s}$ is concerned, it may be said that this dependance is small, at least up to a value of the angle of incidence of about 50° ÷ 60°; this is shown in the fig. 7 in which are represented the variation laws of

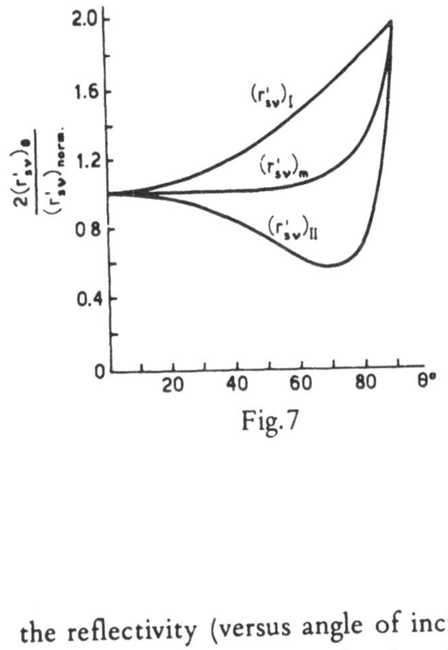

Fig.7

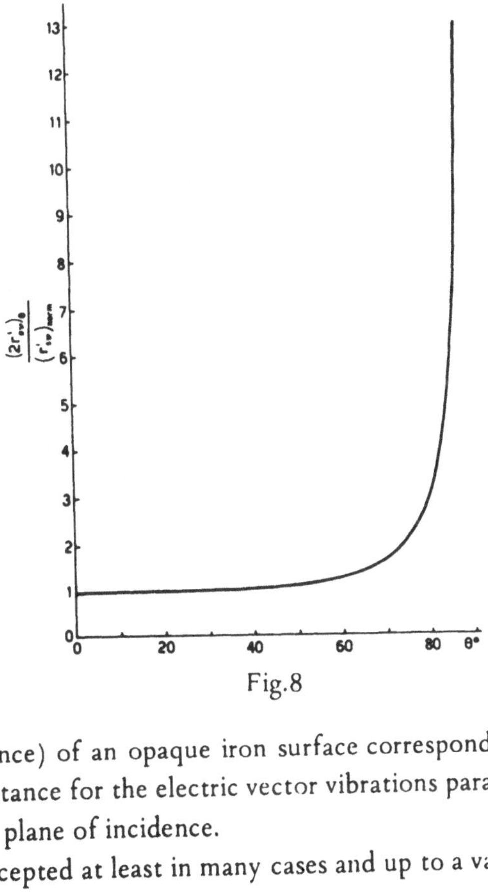

Fig.8

the reflectivity (versus angle of incidence) of an opaque iron surface corresponding to the components of amplitude reflectance for the electric vector vibrations parallel and perpendicular, respectively, to the plane of incidence.

A relation that can be accepted at least in many cases and up to a value of $\vartheta \cong 80°$ to represent such a dependance is

$$(r'_{s\nu})_{\vartheta} = \frac{(r'_{s\nu})\ \text{normal}}{2} \left(\cos\vartheta + \frac{1}{\cos\vartheta}\right) \qquad (10.25)$$

The diagram of fig. 8 gives the variation law of $2(r'_{s\nu})_{\Theta} / (r'_{s\nu})_{\text{norm}}$ corresponding to the eq. (10.25).

10.d. Determination of the biangular reflectance.

When the relation, which gives $r'_{s\nu}$ as a function of ν and ϑ, is known either by theoretical or experimental researches, one can obtain $r^*_{s\nu}$ solving for $r^*_{s\nu}$ the eq. (10.12'), under the assumption that $r^*_{s\nu}$ is a function of the angle Θ of the reflected and incident rays only.

Take

(10.26)
$$r^*_{s\nu} = \sum_{\ell=0}^{\infty} a_{\ell\nu} P_\ell (\cos \Theta)$$

being P_ℓ the Legendre polynomial of degree l, while it is

$$\cos(\Theta) = \cos\vartheta \cos\vartheta' + \sin\vartheta \sin\vartheta' \cos(\varphi-\varphi') =$$

(10.27)
$$= \mu\mu' + (1-\mu^2)^{1/2}(1-\mu'^2)^{1/2} \cdot \cos(\varphi-\varphi')$$

Expanding the Legendre polynomials for the argument $(\cos\Theta)$ by the addition theorem one finds

$$\sum_{\ell=0}^{\infty} a_{\ell\nu} P_\ell(\cos\Theta) = \sum_{\ell=0}^{\infty} \Big\{ a_{\ell\nu} P_\ell(\mu) P_\ell(\mu') + 2 \sum_{m=1}^{\ell} \frac{(\ell-m)!}{(\ell+m)!} P_\ell^m(\mu) P_\ell^m(\mu') \cdot$$
$$\cdot \cos m (\varphi' - \varphi) \Big\}$$

where the P_ℓ^m are the associated Legendre functions (or spherical functions) of the first kind. Inverting the order of summation of the right-hand side of the above eq. (Chandrasekhar [17]) one finds

$$\sum_{\ell=0}^{\infty} a_{\ell\nu} P_\ell \cos\Theta = \sum_{m=0}^{\infty} (2-\delta_{0,m}) \Big\{ \sum_{\ell=m}^{\infty} a_{\ell\nu}^{(m)} P_\ell^m(\mu) P_\ell^m(\mu') \Big\} \cos m (\varphi' - \varphi)$$

being

$$a_{\ell\nu}^{(m)} = a_{\ell\nu} \frac{(\ell-m)!}{(\ell+m)!} \quad (\ell = m, m+1, m+2, \ldots.)$$

and $\delta_{0,m} = 1$, if $m = 0$; $\delta_{0,m} = 0$ if $m \neq 0$,

Thus, it turns out that

$$r'_{s\nu}(\mu) = \sum_{m=0}^{\infty} (2-\delta_{o,m}) \left\{ \sum_{\ell=m}^{\infty} a_{\ell\nu} \overset{(m)}{P}{}^{m}_{\ell}(\mu) \int_0^1 \mu' \overset{m}{P}_{\ell}(\mu') \right\} \int_0^{2\pi} d\varphi' \cos m(\varphi' - \varphi)\, d\varphi =$$

$$= \sum_{\ell=0}^{\infty} 2\pi\, a_{\ell\nu} P_{\ell}(\mu) \int_0^1 \mu' P_{\ell}(\mu')\, d\mu' \qquad (10.28)$$

since

$$\int_0^{2\pi} \cos m(\varphi' - \varphi)\, d\varphi' = 0 \quad \text{if} \quad m \neq 0$$
$$= 2\pi \quad \text{if} \quad m = 0$$

and therefore

$$r'_{s\nu}(\mu) = \sum_{\ell=0}^{\infty} A_{\ell\nu} P_{\ell}(\mu) \qquad (10.29)$$

being

$$A_{\ell\nu} = 2\pi a_{\ell\nu} \int_0^1 \mu' P_{\ell}(\mu')\, d\mu'$$

Multiply by $P_n(\mu)$ both side of the eq.° (10.29) and integrate within the limits $(-1) \div (1)$; writing

$$\int_{-1}^{1} r'_{s\nu}(\mu) P_n(\mu)\, d\mu = B_{n\cdot\mu} \qquad (10.30)$$

one obtains

$$B_{\ell\nu} = A_{\ell\nu} \frac{2}{2\ell + 1} \qquad (10.31)$$

$$a_{\ell\nu} = \frac{2\ell + 1}{4\pi} \frac{B_{\ell\nu}}{\int_0^1 \mu' P_{\ell}(\mu')\, d\mu'} \qquad (10.32)$$

Since $r'_{s\nu}$ has to be a continuous finite function of μ, the series $\sum_0^{\infty} a_{\ell\nu} P_e(\mu)$ results to be certainly convergent and it allows to determine $r''_{s\nu}$ if $r'_{s\nu}$ is a known function.

10.e. Electric field near a surface of a metal.

But it must be observed right now that the electric field, that has been neglected in the general equations is no more negligible near the wall as consequence of the photoelectric emission of the electrons: in effect, the electric field on the surface of the metal is not null, because of the electric charges on this surface, and the mean charge density on the flow field is also not null because of the difference between the numbers of density of the electrons and the ions.

Let be Φ^* the potential of the electric field; it has to satisfy the Poisson's equation

$$\text{div grad } \Phi^* = -4\pi (N_i - N_e) e \tag{10.33}$$

while on the surface Σ it has to be

$$\Phi^* = \text{const.} \tag{10.34}$$

and a similar relation holds at infinity. On the other hand also the assumption that the macroscopic mean velocity of the particles of any type has the same value for all the particles is no more acceptable: thus the continuity equation for the component α_i of the mixture of atoms, ions and electrons takes the form (in the stationary case) (see eq. 4.1)

$$\text{div } (\vec{V}_{\alpha i} N_{\alpha i}) = (M_c)_{\alpha i} + (M_r)_{\alpha i} \tag{10.35}$$

Writing this eq. for the ions and electrons, and observing that the source functions at the right-hand side of the eq. (10.35) are the same in both cases, one finds

$$\text{div} [e (\vec{V}_i N_i - \vec{V}_e N_e)] = 0 \tag{10.36}$$

Now, the vector $[e(\vec{V}_i N_i - \vec{V}_e N_e)]$ is the electric current vector $\vec{j}$: in the case under consideration, in which all the surfaces that limit the field are insulated, so that in each of them the eq. (10.34) holds, it turns out that

$$\vec{V}_i N_i = \vec{V}_e N_e \tag{10.37}$$

One can express $\vec{V}_{\alpha i}$ by means of the rel. (4.3), where the diffusion velocity $\vec{u}_{\alpha i}$ of the species α_i is given by the Chapman-Cowling's formula

$$\vec{u}_{\alpha i} = \frac{N_o^2}{\rho N_{\alpha i}} \Sigma_\beta m_\beta D_{\alpha i,\beta} \vec{d}_\beta - \frac{1}{m_{\alpha i} N_{\alpha i}} D^{(T)}_{\alpha i} \text{ grad log } T_{\alpha i} \quad (\beta \neq \alpha i) \tag{10.38}$$

wherein $N_o = \Sigma_{\alpha i} N_{\alpha i}$ is the number of density of the particles of the mixture; m_β is the mass of the particle β; $D_{\alpha i,\beta}$ (which has the physical dimensions of (length / (time)) is the multicomponent diffusion coefficient for diffusion of the component α_i through component β in the presence of all the other components; $\vec{d}_\beta$ is the vector

$$\vec{d}_\beta = \text{grad}\left(\frac{N_{\alpha i}}{N_o}\right) + \left(\frac{N_{\alpha i}}{N_o} - \frac{m_{\alpha i} N_{\alpha i}}{\rho}\right) \text{grad} \log p_{\alpha i} ; \qquad (10.39)$$

finally $D_{\alpha i}^{(T)}$(which has the dimensions of (mass)/(length) x (time) is the multi-component thermal diffusion coefficient for the component α_i in the mixture.

It appears therefore that the first term at the right-side of the eq. (10.38) defines the concentration and pressure diffusion while the second term gives the thermal diffusion.

The equations which express the conservation of momentum and energy for each species are not given here, but it is clear that in these equations as well as in the eq. (10.35) the terms which give the influence of the electric field have the same form as in the case of a non radiating gas.

11. Simple applications.

In order to show how the general equations above indicated are applied in radiative gasdynamics, two of the simplest problems are now considered; each of them involves one only transition between two states, and besides it will be assumed that the temperature of the electrons is equal to that of atoms and ions.

11.a. Excitation to the first energy level from the ground level.

The expression of the absorption coefficient κ_ν reduces now to the form

$$\kappa_\nu = N_1 \sigma_{12} - N_2 \sigma_{21} = \sigma_{12}\left(N_1 - N_2 \frac{g_1}{g_2}\right) \qquad (11.1)$$

being N_1, N_2 respectively the numbers of density of the atoms at the ground state and the first excited level, and using the rel. (6.24). On the other hand the rel. (6.18) gives

$$\sigma_{12} = \kappa_o \Phi_\nu \qquad (11.2)$$

if

$$\kappa_o = \frac{\pi e^2}{m_e c} \frac{f_{12}}{\nu_{21}} \quad ; \quad \Phi_\nu = \frac{1}{\pi} \frac{\delta/\nu_{21}}{\left(\frac{\nu}{\nu_{21}} - 1\right)^2 + \left(\frac{\delta}{\nu_{21}}\right)^2} \quad . \qquad (11.3)$$

Thus, from the expression of the coefficient φ given by (5.10) it turns out that

$$\varphi = \frac{1 - \exp(-h\nu_{21}/kT)}{\frac{N_1}{N_2}\frac{\bar{N}_2}{\bar{N}_1} - \exp(-h\nu_{21}/kT)} \cong \frac{N_2}{N_1}\frac{\bar{N}_1}{\bar{N}_2} \tag{11.4}$$

if $kT \ll h\nu_{21}$.

The equation of the radiative energy transfer becomes

$$\vec{s}\cdot\text{grad}\, I_\nu(\vec{s}) = -\kappa_o\Phi_\nu\left(N_1 - N_2\frac{g_1}{g_2}\right)\left[I_\nu(\vec{s}) - \frac{N_2}{N_1}\frac{\bar{N}_1}{\bar{N}_2}B_\nu\right] \tag{11.5}$$

Writing $N_0 = N_1 + N_2$, so that N_o represents the number of density of atoms at any level, and denoting by α_2 the excitation degree, defined by the rel. $\alpha_2 = N_2/N_0$, one finds:

$$N_2 = \alpha_2 N_o \quad ; \quad N_1 = (1-\alpha_2)N_o \tag{11.6}$$

so that the eq. (11.5) becomes

$$\vec{s}\cdot\text{grad}\; I_\nu(\vec{s}) = -\kappa_o\Phi_\nu N_o\left[1-\alpha_2\left(1+\frac{g_1}{g_2}\right)\right]\left[I_\nu(s) - \frac{\alpha_2}{1-\alpha_2}\frac{1-\bar{\alpha}_2}{\bar{\alpha}_2}B_\nu\right] \tag{11.7}$$

where $\bar{\alpha}_2$ is the value of α_2 in condition of thermodynamic equilibrium. As for the dependence of the physical quantities on the spatial coordinates, suppose that they depend only on the coordinate x: the eq. (11.7) can be written under the form

$$\cos\vartheta\,\frac{dI_\nu}{dx} = -\kappa_o N_o\Phi_\nu\left[1-\alpha_2(1+g_1/g_2)\right]\left[I_\nu - \frac{\alpha_2}{1-\alpha_2}\frac{1-\bar{\alpha}_2}{\bar{\alpha}_2}B_\nu\right] \tag{11.8}$$

being ϑ the angle that the unit vector $\vec{s}$ forms with the x axis. Let be

$$d\eta = \kappa_o N_o[1-\alpha_2(1+g_1/g_2)]dx = \kappa_o^*\rho[1-\alpha_2(1+g_1/g_2)]dx = \frac{dx}{\ell} \tag{11.9}$$

where ρ is the density of the gas; $\kappa_o^* = \kappa_o m_a^{-1}$, so that κ_o^* has the meaning of absorption coefficient, integrated over all frequencies, per unit mass: l, finally, has the meaning of "photon a mean path", namely of the penetration length of the radiation. The eq. (11.8) reduces to the

$$\cos\vartheta \frac{dI_\nu}{d\eta} = -\Phi_\nu\left(I_\nu - \frac{\alpha_2}{1-\alpha_2}\frac{1-\bar{\alpha}_2}{\bar{\alpha}_2} B_\nu\right) \tag{11.10}$$

Assuming that the radiation field extends to infinity in both direction, so that η can have any value between $-\infty$ and $+\infty$ from the boundary condition $I_\nu = B_\nu$ for $\eta = \pm\infty$, one deduces

$$I_\nu(\vartheta,\eta) = \int_{-\infty}^{0} \Phi_\nu \frac{\alpha_2}{1-\alpha_2}\frac{1-\bar{\alpha}_2}{\bar{\alpha}_2} B_\nu \exp(\Phi_\nu \chi)\, d\chi \tag{11.11}$$

being $\chi = \dfrac{\eta'-\eta}{\mu}$; $\mu = \cos\vartheta$

The radiative energy flux through the unit area perpendicular to the x axis is given by the rel.

$$F_{\nu x} = 2\pi\left[\int_0^{\pi/2} I_\nu \cos\vartheta \sin\vartheta\, d\vartheta - \int_0^{\pi/2} I_\nu \cos\vartheta' \sin\vartheta'\, d\vartheta'\right] \tag{11.12}$$

if $\vartheta' = \vartheta - \pi$, so that one obtains

$$F_{\nu x} = 2\pi\int_{-\infty}^{\infty} \operatorname{sign}(\eta-\eta')\Phi_\nu \frac{\alpha_2}{1-\alpha_2}\frac{1-\bar{\alpha}_2}{\bar{\alpha}_2} B_\nu E_\nu[\Phi_\nu|\eta-\eta'|]\, d\eta' \tag{11.13}$$

whereas $E_2(t>0)$ is the exponential integral function of second order defined by the rel.

$$E_n(t>0) = \int_0^1 \mu^{n-2} \exp(-t/\mu)\, d\mu \qquad (n = 1,2,\ldots) \tag{11.14}$$

The radiative energy flux integrated over all frequencies can be obtained by the

$$F_x = \int_0^\infty F_{\nu x}\, d\nu = 2\pi\nu_{21}\int_{-\infty}^{\infty} \operatorname{sign}(\eta-\eta')\frac{\alpha_2}{1-\alpha_2}\frac{1-\bar{\alpha}_2}{\bar{\alpha}_2} B_{\nu_{21}} G(|\eta-\eta'|)\, d\eta' \tag{11.15}$$

where

$$G(|\eta-\eta'|) = \int_0^\infty \Phi_\nu \frac{B_\nu}{B_{\nu_{21}}} E_2(\Phi_\nu|\eta-\eta'|)\, d(\nu/\nu_{21}) \tag{11.16}$$

An investigation of the integral at the right side of the rel. (11.16) shows that, if one neglects the terms of the order of magnitude as $(\delta/\nu_{21})^{1/2}$ (and therefore of about 10^{-4}), the value of the function G can be taken equal to zero if $|\eta-\eta'| > (\delta/\nu_{21})^{1/2}$, while it is about one if $|\eta-\eta'| \to 0$.

Consequently one can take

$$F_x = 2\pi\nu_{21}\left[\int_{\eta-\left(\frac{\delta}{\nu_{21}}\right)^{1/2}}^{\eta} \frac{\alpha_2}{1-\alpha_2}\,\frac{1-\bar{\alpha}_2}{\bar{\alpha}_2}\,B_{\nu_{21}}\,d\eta' - \int_{\eta}^{\eta+\left(\frac{\delta}{\nu_{21}}\right)^{1/2}} \frac{\alpha_2}{1-\alpha_2}\,\frac{1-\bar{\alpha}_2}{\bar{\alpha}_2}\,B_{\nu_{21}}\,d\eta'\right]$$

(11.17)

If in the small interval in which η varies, the magnitude of $Y=[\alpha_2/(1-\alpha_2)]\,[(1-\bar{\alpha}_2)/(\bar{\alpha}_2)]\,B_{\nu_{21}}$ is assumed to vary slowly enough in order that one can write

$$Y = (Y)_{\eta'=\eta} + \left(\frac{dY}{d\eta'}\right)_{\eta'=\eta}(\eta'-\eta) = Y_o + \dot{Y}_o(\eta'-\eta)$$

one finds

(11.18) $$F_x = -\,2\pi\nu_{21}\left(\frac{\delta}{\nu_{21}}\right)\,\ell\,\frac{d}{dx}\left(\frac{\alpha_2}{1-\alpha_2}\,\frac{1-\bar{\alpha}_2}{\bar{\alpha}_2}\,B_{\nu_{21}}\right)$$

that in the case of thermodynamic equilibrium is corresponding to the Rosseland approximation.

In analogous way one can evaluate the radiative source function. It is

(11.19) $$(M_r)_{12} = -\,N_1 B_{12}\left(1-\frac{N_2}{N_1}\,\frac{g_1}{g_2}\right)\left(C_{\nu_{21}}\int_0^{\infty}\Phi_\nu J_\nu\, d\nu/\nu_{21} - \frac{N_2}{N_1}\,\frac{\bar{N}_1}{\bar{N}_2}\,B_{\nu_{21}}\right)$$

Since

$$B_{12} = \frac{4\pi e^2}{m_e c}\,f_{12}\,\frac{1}{h\nu_{21}} = 4\pi\,\frac{\kappa_o}{h}$$

it is also

$$(M_r)_{12} = -\,N_o[\,1-\alpha_2(1+g_1/g_2)]\,\frac{4\pi\kappa_o}{h}\,B_{\nu_{21}}\left(C_{\nu_{21}}\int_0^{\infty}\Phi_\nu\,\frac{J_\nu}{B_{\nu_{21}}}\,d\nu/\nu_{21} - \frac{\alpha_2}{1-\alpha_2}\,\frac{1-\bar{\alpha}_2}{\bar{\alpha}_2}\right)$$

(11.20)

On the other hand if one multiplies the eq. (11.7) by $d\Omega$ and one integrates over all directions one finds

$$\operatorname{div}\vec{F}_\nu = -4\pi\kappa_o N_o \Phi_\nu B_{\nu_{21}}\left[1-\alpha_2\left(1+\frac{g_1}{g_2}\right)\right]\left(\frac{J_\nu}{B_{\nu_{21}}} - \frac{\alpha_2}{1-\alpha_2}\frac{1-\bar{\alpha}_2}{\bar{\alpha}_2}\frac{B_\nu}{B_{\nu_{21}}}\right) \tag{11.21}$$

Dividing by $(h\nu_{21})$ and integrating over all frequencies one obtains

$$\operatorname{div}\left(\int_0^\infty (h\nu_{21})^{-1}\vec{F}_\nu d\nu\right) = -\frac{4\pi\kappa_o}{h}\frac{B_{\nu_{21}}}{C_{\nu_{21}}}\left[1-\alpha_2(1+g_1/g_2)\right]\left[C_{\nu_{21}}\cdot\right.$$
$$\left.\cdot\int_0^\infty \Phi_\nu \frac{J_\nu}{B_{\nu_{21}}} d\nu/\nu_{21} - \frac{\alpha_2}{1-\alpha_2}\frac{1-\bar{\alpha}_2}{\bar{\alpha}_2}\right)\Bigg] \tag{11.22}$$

By comparison with the expression of $(M_r)_{12}$ one finds

$$(M_r)_{12} = C_{\nu_{21}}\operatorname{div}\left[\int_0^\infty (h\nu_{21})^{-1}\vec{F}_\nu d\nu\right] \tag{11.23}$$

and, in the case of dependence of all quantities only on the x coordinate,

$$(M_r)_{12} = \frac{C_{\nu_{21}}}{\ell}\frac{d}{d\eta}(h\nu_{21})^{-1}\,2\pi\nu_{21}\int_0^\infty \operatorname{sign}(\eta-\eta')\frac{\alpha_2}{1-\alpha_2}\frac{1-\bar{\alpha}_2}{\bar{\alpha}_2}B_{\nu_{21}}G(|\eta-\eta'|)d\eta' \tag{11.24}$$

Neglecting the diffusion effect, and therefore taking $\vec{V}_1 = \vec{V}_2 = \vec{U}$, one finds

$$\operatorname{div}[\rho m_a^{-1}\vec{U}(1-\alpha_2)] = -\operatorname{div}[\rho m_a^{-1}\vec{U}\alpha_2]$$

from which

$$m_a^{-1}\rho U\alpha_2 = -C_{\nu_{21}}\frac{2\pi}{h}\int_{-\infty}^{\infty}\operatorname{sign}(\eta-\eta')\frac{\alpha_2}{1-\alpha_2}\frac{1-\bar{\alpha}_2}{\bar{\alpha}_2}B_{\nu_{21}}G(|\eta-\eta'|)d\eta' -$$
$$-\ell\int_{-\infty}^{\eta}(M_c)_{12}d\eta' \tag{11.25}$$

being $(M_c)_{12}$ the collisional source function corresponding to the collisional excitation. The rel. (11.25) defines the variation law of α_2.

One can apply the eq. (11.25) to the determination of the precursor effect of the radiation in the case of a strong normal shock propagating in the x direction with a constant velocity in a gas, wich at infinity upward of the shock is at rest. Referring the flow to a coordinate system bound to the shock front, the conditions of the flow are just those which have been considered in this section.

The eq. (11.25), when one takes as U the velocity propagation of the shock, and as ρ the density $\rho_{-\infty}$ at infinity upward, can be written under the form

$$m_a^{-1}\ \rho_{-\infty} U\alpha_2 = -\ C_{\nu_{21}} \frac{2\pi}{h} \int_{-\infty}^{\infty} \mathrm{sign}(\eta - \eta')\ \frac{\alpha_2}{1-\alpha_2}\ \frac{1-\bar{\alpha}_2}{\bar{\alpha}_2}\ B_{\nu_{21}}\, d\eta'$$

(11.26)

$$\cdot \int_0^{\infty} \Phi_\nu \frac{B_\nu}{B_{\nu_{21}}}\ E_2(\Phi_\nu |\eta - \eta'|)\ d\,\nu/\nu_{21} - \ell \int_{-\infty}^{\eta} (M_c)_{12}\, d\eta'$$

Suppose that the temperature upward of the shock is small enough that the emission and the collisional processes are negligible in that region; besides assume that for $\eta'>0$ (namely, downward of the shock) one can take $\alpha_2=\bar{\alpha}_2$ and the temperature T constant and equal to the value corresponding to the equilibrium conditions at infinity downward (hypothesis that can be accepted when $\eta<0$ and $/\eta/ \gg 1$), the eq. (11.26) becomes

$$m_a^{-1}\ \rho_{-\infty} U\alpha_2 = C_{\nu_{21}} \frac{2\pi}{h} B_{\nu_{21}} \int_0^{\infty} \Phi_\nu \frac{B_\nu}{B_{\nu_{21}}}\ d\,\nu/\nu_{21} \int_0^{\infty} E_2(\Phi_\nu |\eta - \eta'|)\, d\eta'$$

and since

$$\int_0^{\infty} E_2(\Phi_\nu |\eta - \eta'|)\ d\eta' = \frac{1}{\Phi_\nu} E_3(-\Phi_\nu \eta) \tag{11.27}$$

one finds

$$m_a^{-1}\ \rho_{-\infty} U\alpha_2 = C_{\nu_{21}} \frac{2\pi}{h} B_{\nu_{21}} \int_0^{\infty} \frac{B_\nu}{B_{\nu_{21}}}\ E_3(-\Phi_\nu \eta)\, d\,\nu/\nu_{21} \tag{11.28}$$

If one supposes that the eq. (11.28) can be applied even when $\eta=0$, being $E_3(0)=1/2$, one deduces

$$m_a^{-1}\ \rho_{-\infty} U\alpha_2(0) = C_{\nu_{21}} \frac{2\pi}{h} B_{\nu_{21}}\ 1/2 \int_0^{\infty} \frac{B_\nu}{B_{\nu_{21}}}\ d\,\nu/\nu_{21} \tag{11.29}$$

Thus it is

$$\frac{\alpha_2(\eta<0)}{\alpha_2(0)} = 2\ \frac{\int_0^{-\infty} \frac{B_\nu}{B_{\nu_{21}}}\ E_3(-\Phi_\nu \eta)\, d\,\nu/\nu_{21}}{\int_0^{\infty} \frac{B_\nu}{B_{\nu_{21}}}\ d\,\nu/\nu_{21}} \tag{11.30}$$

Since in almost all the intervals over which the integral is extended the function $E_3(-\Phi_\nu \eta)$, for $|\eta|$ large but finite (f.i. such that $|\eta|(\delta\nu_{21})^{1/3} \ll 1$), is very close to 1/2, it appears from the rel. (11.30)) that for a large interval of values of η, starting from $\eta=0$ and $\eta<0$, the quantity α_2 decreases very slowly, and therefore the precursor effect extends to a very large distance. It must, however, be observed that, at least in the experimental conditions, the region downward of the shock is not infinite: if H is the thickness of this region one should write instead of the rel. (11.27) the rel.

$$\int_0^H E_2(\Phi_\nu |\eta-\eta'|)d\eta' = \frac{1}{\Phi_\nu}\left\{E_3(-\Phi_\nu\eta) - E_3[\Phi_\nu(H-\eta)]\right\} \quad (11.31)$$

and consequently one should find

$$m_a^{-1}\rho_{-\infty} U\alpha_2 = C_{\nu_{21}}\frac{2\pi}{h}B_{\nu_{21}}\int_0^\infty \frac{B_\nu}{B_{\nu_{21}}}\left|E_3(-\Phi_\nu\eta) - E_3[\Phi_\nu(H-\eta)]\right| d\frac{\nu}{\nu_{21}}$$

$$m_a^{-1}\rho_{-\infty} U\alpha_2(0) = C_{\nu_{21}}\frac{2\pi}{h}B_{\nu_{21}}\int_0^\infty \frac{B_\nu}{B_{\nu_{21}}}[1/2 - E_3(\Phi_\nu H)]d\,\nu/\nu_{21} \quad (11.32)$$

For $H(\delta/\nu_{21})^{1/3} \ll 1$ it is also $E_3(\Phi_\nu H) \cong 1/2$, so that the precursor effect would be extended always to a large distance, but it should be very small.

11.b. Ionization from the ground state.

In this case since

$$\frac{\bar{\sigma}_{21}}{\sigma_{21}} = \frac{\bar{N}_2}{N_2} \quad ; \quad \frac{\bar{\sigma}_{21}}{\bar{\sigma}_{12}} = \frac{\bar{N}_1}{\bar{N}_2}\exp(-h\nu/kT)$$

one finds

$$\kappa_\nu = N_1\sigma_{12}\left[1 - \frac{N_2^2}{N_1}\frac{\bar{N}_1}{\bar{N}_2^2}\exp(-h\nu/kT)\right]$$

$$\varphi = \frac{\frac{\bar{N}_1}{\bar{N}_2}\frac{\bar{\sigma}_{12}}{\bar{\sigma}_{21}} - 1}{\frac{N_1}{N_2}\frac{\sigma_{12}}{\sigma_{21}} - 1} = \frac{1-\exp(-h\nu/kT)}{\frac{N_1}{N_2^2}\frac{\bar{N}_2^2}{\bar{N}_1} - \exp(-h\nu/kT)}$$

Let be α_j the ionization degree defined as

$$\alpha_j = \frac{N_2}{N_1 + N_2} = \frac{N_2}{N_0}$$

Using the Kramer's formula to express the cross-section σ_{12}, one has

$$\sigma_{12} = \frac{32\pi^2 e^6}{3\sqrt{3}ch^3} \frac{\mathcal{R}_y Z_e^4}{\nu^3} = \frac{32\pi^2 e^6}{3\sqrt{3}ch^3} \frac{\mathcal{R}_y Z_e^4}{\nu_j^3} \left(\frac{\nu_j}{\nu}\right)^3$$

The equation of the radiative energy transfer becomes

$$\vec{s}\cdot\text{grad}\ I_\nu(\vec{s}) = -N_o(1-\alpha_j)\frac{32\pi^2 e^6}{3\sqrt{3}ch^3}\frac{\mathcal{R}_y Z_e^4}{\nu_j^3}\left(\frac{\nu_j}{\nu}\right)^3$$

$$\cdot\left[1 - \frac{\alpha_j^2}{\bar{\alpha}_j^2}\frac{1-\bar{\alpha}_j}{1-\alpha_j}\exp(-h\nu/kT)\right]\left[I_\nu - B_\nu \frac{1-\exp(-h\nu/kT)}{\frac{1-\alpha_j}{\alpha_j^2}\frac{\bar{\alpha}_j^2}{1-\bar{\alpha}_j} - \exp(-h\nu/kT)}\right]$$

Writing

$$N_o = \rho m_a^{-1} \quad ; \quad \kappa = m_a^{-1}\frac{32\pi^2 e^6}{3\sqrt{3}\ ch^3}\frac{\mathcal{R}_y Z_e^4}{\nu_j^3}$$

and neglecting the terms of the order of magnitude as $\exp(-h\nu/kT)$ with respect to the unity one finds, under the same assumption as in 11.a.

$$\cos\vartheta\frac{dI_\nu}{dx} = -\rho(1-\alpha_j)\kappa\left(\frac{\nu_j}{\nu}\right)^3\left[I_\nu - \frac{\alpha_j^2}{1-\alpha_j}\frac{1-\bar{\alpha}_j}{\bar{\alpha}_j^2}B_\nu\right]$$

Let be

$$d\eta = \kappa\rho(1-\alpha_j)dx = \frac{dx}{\ell_j}$$

and therefore

$$\eta = \int^x \kappa\rho(1-\alpha_j)\ dx$$

where η has the meaning of optical distance, while lj has again, as in 11.a., the meaning of photon mean free path (for frequency ν_j).

One can write

$$\cos\vartheta \frac{dI_\nu}{d\eta} = -\left(\frac{\nu_j}{\nu}\right)^3 \left[I_\nu - \frac{\alpha_j^2}{1-\alpha_j}\frac{1-\bar{\alpha}_j}{\bar{\alpha}_j^2} B_\nu\right]$$

In any case in which η can have any value between $-\infty \div +\infty$, one can obtain

$$I_\nu(\vartheta,\eta) = \int_{-\infty}^{0} \left(\frac{\nu_j}{\nu}\right)^3 \frac{\alpha_j^2}{1-\alpha_j}\frac{1-\bar{\alpha}_j}{\bar{\alpha}_j^2} B_\nu \exp\left[\left(\frac{\nu_j}{\nu}\right)^3 X\right] dX$$

where

$$X = \frac{\eta-\eta'}{\mu} \; ; \; \mu = \cos\vartheta$$

The correspondent expression of the radiative energy flux $F_{\nu x}$ is now

$$F_{\nu x} = 2\pi \int_{-\infty}^{\infty} \mathrm{sign}\,(\eta-\eta') \frac{\alpha_j^2}{1-\alpha_j}\frac{1-\bar{\alpha}_j}{\bar{\alpha}_j^2} \left(\frac{\nu_j}{\nu}\right)^3 B_\nu E_2\left[\left(\frac{\nu_j}{\nu}\right)^3 |\eta-\eta'|\right] d\eta'$$

while the integrated flux F_x

$$F_x = 4\pi\nu_j^4 \frac{h}{c^2} \int_{-\infty}^{\infty} \mathrm{sign}\,(\eta-\eta') \frac{\alpha_j^2}{1-\alpha_j}\frac{1-\bar{\alpha}_j}{\bar{\alpha}_j^2} G\,(|\eta-\eta'|)\, d\eta'$$

being

$$G\,(|\eta-\eta'|) = \int_1^{\infty} \exp\left[-(T_j/T)\frac{\nu}{\nu_j}\right] E_2\left[\left(\frac{\nu_j}{\nu}\right)^3 |\eta-\eta'|\right] d\nu/\nu_j$$

being Tj the ionization temperature defined by the rel. $T_j = \epsilon_j/k$. As for the radiative source function one finds now

$$(M_r)_{ion.} = -4\pi N_1 \int_{\nu_j}^{\infty} (h\nu)^{-1}\sigma_{12}\left(J_\nu - \frac{N_2^2}{N_1}\frac{\bar{N}_1}{\bar{N}_2^2} B_\nu\right) d\nu = \mathrm{div}\int_{\nu_j}^{\infty} (h\nu)^{-1}\vec{F}_\nu\, d\nu$$

as one could find with a procedure perfectly analogous to that indicated in the case (11.a.). If the quantities depend only on the x one has

$$(M_r)_{ion} = \frac{2\pi}{h\ell_j}\frac{d}{d\eta}\int_{-\infty}^{\infty} \mathrm{sign}(\eta-\eta')\, B_{\nu_j} \frac{\alpha_j^2}{1-\alpha_j}\frac{1-\bar{\alpha}_j}{\bar{\alpha}_j^2} G^*(|\eta-\eta'|)\, d\eta'$$

being
$$B_{\nu_j} = \frac{2h\nu_j^3}{c^2} \exp(-T_j/T)$$

$$G^* = \exp(T_j/T) \int_1^\infty \frac{\nu_j}{\nu} \exp\left[\left(1 - \frac{T_j}{T}\right)\left(\frac{\nu}{\nu_j}\right)\right] E_2\left[\left(\frac{\nu_j}{\nu}\right)^3 |\eta - \eta'|\right] d\,\nu/\nu_j$$

Thus one deduces
$$m_a^{-1} \rho\, U\, \alpha_j = -\frac{2\pi}{h} \int_{-\infty}^{\infty} \mathrm{sign}(\eta - \eta')\, B_{\nu_j} \frac{\alpha_j^2}{1-\alpha_j} \frac{1-\bar{\alpha}_j}{\bar{\alpha}_j^2} G^*(|\eta - \eta'|)\, d\eta' -$$
$$- \ell_j \int_{-\infty}^{\eta} (M_c)_{ion.}\, d\eta'$$

Applying this eq. to the problem of the structure of a strong normal shock propagation with a constant velocity in the x direction, and assuming analogous hypothesis as those indicated in the analogous problem considered in (11.a.), one finds for the precursor effect of the shock

$$m_a^{-1} \rho_{-\infty} U \alpha_j = \frac{2\pi}{h} B_{\nu_j} \int_1^\infty \frac{B_\nu}{B_{\nu_j}} \left(\frac{\nu_j}{\nu}\right) E_3\left[-\left(\frac{\nu_j}{\nu}\right)^3 \eta\right] d\,\nu/\nu_j$$

and therefore
$$\frac{\alpha_j(\eta<0)}{\alpha_j(0)} = 2\, \frac{\displaystyle\int_1^\infty \frac{B_\nu}{B_{\nu_j}} \left(\frac{\nu_j}{\nu}\right) E_3\left[-\left(\frac{\nu_j}{\nu}\right)^3 \eta\right] d\,\nu/\nu_j}{\displaystyle\int_1^\infty \frac{B_\nu}{B_{\nu_j}} \left(\frac{\nu_j}{\nu}\right) d\,\nu/\nu_j}$$

which is formally analogous to the rel. (11.30).

It appears, however, that now for a large interval of frequencies, such that $(\nu_j/\nu)^3 |\eta| \gg 1$, if $|\eta| \gg 1$, the function at the numerator at the right side is much less than the function at denominator, and therefore $\alpha_j(\eta<0)/\alpha_j(0)$ decreases rapidly with increasing $|\eta|$.

REFERENCES

[1] J.T. Howe, J.R. Viegas
NASA TR–R 159 (1963).

[2] K.H. Yosikava, B.H. Vick
NASA TN D–1074 (1961).

[3] V. Veisskoff, E. Wigner
Zs. f. Phys. Vol. 63, pag. 54 (1930).

[4] L.H. Aller
Astrophys. The atmosphere of the Sun and Stars -
The Ronald Press Comp. N.Y. (1953), pag. 133

[5] L. Goldberg
Astrophys. J., vol. 82 (1935), vol. 84 (1936).

[6] L.H. Aller
(see 4 pag. 244.

[7] L.H. Aller
see 4 pag. 248.

[8] O. Harris
Astrophys. J., vol. 108 (1948).

[9] D.R. Bates
M.N., vol. 106, pag. 432 (1946).

[10] D.R. Bates, A. Damgaard
Phyl. Trans. Roy. Soc. A., vol. 242 pag. 101 (1959).

[11] L.M. Biberman, G.E. Norman
Opt. Spect. (USSR) (English transl.) vol. 8, pag. 230 (1960).

[12] H.C. Hayden, R.C. Amme
Phys. Rev., vol. 141, pag. 401 (1966).

[13] H.C. Hayden, N.G. Utterbach
Phys. Rev., vol. 135 A., pag. 1575 (1964).

[14] R.V. Dunkle
Theory and Fund. Res. in Heat Transfer, pag. 1-31, edit. by J.A. Clark,
Symp. Publ. Div. Pergamon Press (1963).

[15] G.A.W. Rudgers
Handbuch der Physik, vol. XXVI, pag. 129-170.

[16] J.D. Price Phy
Phys. Soc. London, vol. 62 (1949).

[17] S. Chandrasekhar
Radiation Transfer, Oxford Univ. Press London (1950).

CONTENTS

Printed in the United States
By Bookmasters